Lecture Notes in Applied and Computational Mechanics

Volume 97

This series aims to report new developments in applied and computational mechanics - quickly, informally and at a high level. This includes the fields of fluid, solid and structural mechanics, dynamics and control, and related disciplines. The applied methods can be of analytical, numerical and computational nature. The series scope includes monographs, professional books, selected contributions from specialized conferences or workshops, edited volumes, as well as outstanding advanced textbooks.

Indexed by EI-Compendex, SCOPUS, Zentralblatt Math, Ulrich's, Current Mathematical Publications, Mathematical Reviews and MetaPress.

More information about this series at http://www.springer.com/series/4623

Christian Weißenfels

Simulation of Additive Manufacturing using Meshfree Methods

With Focus on Requirements for an Accurate Solution

Springer

Christian Weißenfels
Institute of Continuum Mechanics
Leibniz University Hannover
Garbsen, Germany

ISSN 1613-7736 ISSN 1860-0816 (electronic)
Lecture Notes in Applied and Computational Mechanics
ISBN 978-3-030-87336-3 ISBN 978-3-030-87337-0 (eBook)
https://doi.org/10.1007/978-3-030-87337-0

This Springer imprint is published by the registered company Springer Nature Switzerland AG
The registered company address is: Gewerbestrasse 11, 6330 Cham, Switzerland

Preface

The big promise of Additive Manufacturing is the rapid production of individualized products directly on-site and on-demand. Annual growth rates are still averaging more than 25%. In recent years, numerous different 3D-printing processes have become established on the market with which various materials can be processed. Nevertheless, there are still several obstacles standing in the way of the breakthrough to a standard manufacturing process. In particular, the speed of fabrication, the quality, and the reproducibility of a printed component still do not meet the requirements of industrial production.

To realize this, all processes during 3D printing must be understood. The influence of individual parameters cannot be estimated based on experiments alone, since the speed of the processes involved is very high. Realistic simulations make it possible to investigate all physical phenomena during Additive Manufacturing. Every single influencing factor can be analyzed on the computer for every fraction of a second. This allows the key factors of 3D printing to be identified.

Unfortunately, the phenomena that occur in Additive Manufacturing pose significant challenges to the simulation. Due to their strong flexibility, meshfree methods have been a big promise for years. Large deformations with free surfaces or the fusion of materials can be easily realized. Meshfree methods are, thus, ideal discretization schemes for Additive Manufacturing simulations. On the other hand, the accuracy of the solution cannot always be guaranteed. Also, the handling strongly depends on the problem. These reasons prevent the use of these schemes in an industrial environment until today.

This monograph provides all requirements on spatial discretization schemes to guarantee an accurate solution of differential equations. In addition, several commonly used meshfree methods are presented, compared, and investigated concerning these requirements. Smoothed Particle Hydrodynamics and the Optimal Transportation Meshfree method are employed to simulate Selective Laser Melting. Peridynamics is used to simulate a deposition process for medical silicone.

The basis for this monograph is my Habilitation thesis, which I wrote during my time at the Institute for Continuum Mechanics at the Leibniz University Hannover. The contents are aimed at graduate students, researchers, and practitioners who want to explore the secrets behind meshfree methods or get into the modeling of Additive Manufacturing processes.

My special thanks go to Prof. Peter Wriggers. He not only gave me the opportunity to do my Ph.D. and Habilitation, but also always gave me his trust and great support over all these years. Professor Wriggers is one of the outstanding scientists in the field of engineering and computational mechanics. Working with him has always been a special honor for me.

I also want to express my gratitude to the Series Editors for inviting me to contribute to this distinguished series from Springer. Along this line, I would also like to thank Springer for a good collaboration.

Since this monograph relies on my Habilitation thesis, I would like to express my appreciation to the reviewers, i.e. Prof. Jörg Schröder and Prof. Dennis Kochmann; the chairmen of the committee Prof. Peter Nyhus and Prof. Jörg Wallaschek; and all other committee members.

In addition, I would like to express my sincere thanks to my colleagues in my research group, Tobias Bode, Jan-Philipp Fürstenau, Philipp Hartmann, and Henning Wessels, who contributed significantly to the content and results in this monograph. Furthermore, I would like to thank Dengpeng Huang, Sandeep Kumar, and Meisam Soleimani for their good cooperation during my Habilitation period.

Hannover, Germany
June 2021

Christian Weißenfels

Contents

Chapter 1
Introduction

The big promise of Additive Manufacturing (AM) is the fast production of individualized products directly on site, on demand, and at low costs. Many nations, such as the USA, Japan, or Great Britain, hope that this technology will lead to their reindustrialization. Countries like China see AM as an opportunity to take the lead in a field of technology (Caviezel et al. 2017). The annual growth rates are on average about 25 percent since the records in 1998 (Wohlers Associates 2017). Revenues from 3D printers, materials, software and accessories, as well as AM services, totaled \$6.063 billion in 2016. The big potential is attributed to Additive Manufacturing in the field of biomedical technology. In 2013, the sales turnover of medical applications amounted to only 1.6% of the total revenue. However, forecasts expect an increase to 21% within the next few years (Ventola 2014). Patient-specific implants, prostheses, and active pharmaceutical ingredients can become the standard in biomedical technology through Additive Manufacturing. The idea of printing human organs is also not a utopia and could in the future remedy the lack of donor organs.

In 1986, Charles Hull was the first to file a patent for a 3D printer (Hull 1986). He was able to print a component layer by layer from a CAD file and called his method stereolithography. The .stl format for importing CAD data was the abbreviation for this technology. In the following years, numerous different printing technologies were established on the market. Nevertheless, there are still several shortcomings that hinder the breakthrough to a large-scale production process. In particular, the speed of the manufacturing process, the quality, and the reproducibility of a printed component, still do not meet the requirements of industrial manufacturing. To achieve this, all phenomena during the printing process must be comprehensively understood. This is the only way to ensure that the material properties, as well as the shape of the printed part, correspond to the specifications planned on the computer beforehand.

In Selective Laser Melting, around 130 parameters influence the printing process (Yadroitsev 2009). The extremely rapid sequence of individual physical phenomena

C. Weißenfels, *Simulation of Additive Manufacturing using Meshfree Methods*,
Lecture Notes in Applied and Computational Mechanics 97,
https://doi.org/10.1007/978-3-030-87337-0_1

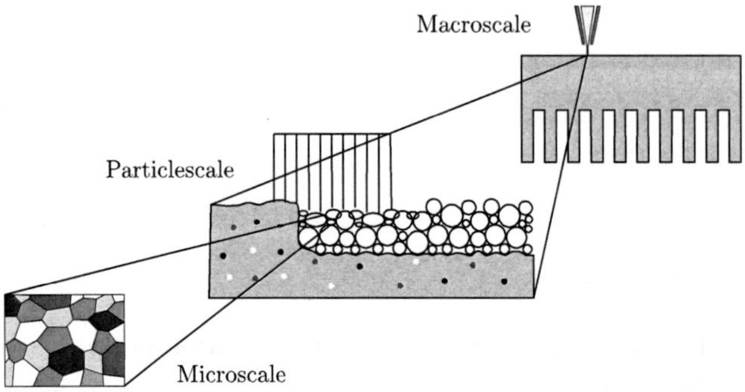

Fig. 1.1 Consideration of different scales when modeling Additive Manufacturing

makes analysis based on experimental evaluations difficult. Realistic simulations on the computer enable the reproduction of all physical processes during Additive Manufacturing. The influence of each phenomenon can be individually analyzed for each fraction of a second which helps to identify the key factors of the printing process. This can dramatically accelerate the optimization of existing 3D printers or pave the way for the development of new printing technologies. Hence, the simulation of Additive Manufacturing is a big driver for the development of next-generation 3D printers.

In addition, realistic simulation data can be directly integrated into the manufacturing process. The adjustment of the process parameters that ensure an optimal print can be based directly on this data.

Simulation means modeling phenomena and usually leads to differential equations. For accurate and efficient modeling, it is often suitable to consider the effects at the individual scales. Different scales can also be defined in the simulation of additive manufacturing, see Fig. 1.1. In most cases, an analytical solution is no longer possible, and mostly numerical schemes are used. In the first step, the differential equation is converted into an algebraic equation employing a spatial and temporal discretization. For each time step, the solution of a system of linear equations provides the approximate solution.

The Finite Element Method (FEM) is currently the most widely used discretization scheme. It is based on a unique subdivision of the domain into a finite number of sections. This scheme convinces by an accurate solution of the differential equation. Due to the fixed subdivision into elements, however, the application of the FEM to simulate Additive Manufacturing processes requires special treatments.

Due to their flexibility, meshfree discretization schemes have been a big promise in computational mechanics for many years. Large deformations with free surfaces or the fusion of materials can be easily realized. Meshfree methods are therefore ideal spatial discretization schemes for the simulation of Additive Manufacturing processes. However, an accurate solution cannot always be guaranteed which prevents the intensive use of these methods by engineers in the industry.

The objective of this work is to provide all requirements on spatial discretization schemes to guarantee an accurate solution of differential equations. In the context of this work, several common meshfree methods are introduced, compared with each other, and investigated concerning the requirements for spatial discretization schemes.

Especially in the field of meshfree methods, there have been numerous developments since the 90s, which have been carried out in many different groups worldwide. This work deals only with the most important developments from the author's point of view. In addition, only selected meshfree methods that have received special attention in recent years are presented. An insight into other schemes can be found e.g. in Li and Liu (2007) or Liu and Liu (2015).

Two Additive Manufacturing processes are simulated employing meshfree discretization schemes. Selective Laser Melting is a laser powder bed fusion process frequently used for printing metallic or ceramic parts. The material properties and the shape of the final product are mainly determined by the behavior of the melt pool. For this reason, the printing of one layer is investigated. The influence of the cooling rate on the micromechanical behavior of the material as well as the influence of printing many layers on the final shape of the part is not considered. Examples of numerical investigations on these scales can be found in Schoinochoritis et al. (2017) and Yan et al. (2018).

The fusion of powder particles is simulated in Körner et al. (2013) using the Lattice-Boltzmann method. Also, Computational Fluid Dynamics (CFD) solvers, which are based on an Eulerian description, are used to analyze this process computationally (Yan et al. 2017). Therein, the surface is detected by the Volume of Fluid method, which was developed in Hirt and Nichols (1981). The coupling of this approach with the Discrete Element Method can be found in Wang et al. (2019). Another option is the use of the Arbitrary Lagrangian–Eulerian (ALE) approach (King et al. 2015).

To simplify the simulation of fusion and fluid flow with free surfaces, the virtual reproduction of the Selective Laser Melting process is carried out using meshfree solution schemes. The Optimal Transportation Meshfree (OTM) method, as well as Smoothed Particle Hydrodynamics (SPH), are employed to discretize the differential equations. Both methods are compared and the outcomes are evaluated. A first simulation using SPH can be found in Russell et al. (2018). However, this approach is limited to a purely 2-dimensional description without considering the influence of recoil pressure and phase change on the mechanical properties of the material. Also, the curvature is not modeled accurately.

The second printing process allows the fabrication of flexible, patient-specific neural implants. Since conventional technologies are not suitable, in Stieghorst (2017) a new approach is developed.

This work contains the following topics:

Chapter 2 gives an introduction to Additive Manufacturing. Besides presenting different 3D printing technologies, the potentials, as well as the current gaps and needs of Additive Manufacturing, are shown.

Chapter 3 provides the equations for the description of 3D printing processes. A special focus is on the effects at the boundary layer. Surface tension, Marangoni convection, and recoil pressure are key factors in Selective Laser Melting. Also, the description of the phase change is central to the accurate reproduction of these processes. Photopolymerization involves a chemical reaction, which can also be described by a differential equation.

Chapter 4 gives an introduction to the numerical solution of differential equations. An important step is the spatial discretization. For an accurate solution, several requirements have to be fulfilled. For this reason, all the necessary conditions are given in detail and a new requirement is introduced.

Chapter 5 deals with meshfree Galerkin schemes. The focus is on the Optimal Transportation Meshfree (OTM) method according to Li et al. (2010). Especially the shortcomings of this method are pointed out. First, another scheme that can be seen as a precursor to the OTM method is briefly presented and discussed.

Chapter 6 introduces Smoothed Particle Hydrodynamics (SPH) which was developed in the seventies by Gingold and Monaghan (1977) and in parallel by Lucy (1977). Over the years, there have been numerous further developments. The main field of application is the simulation of fluid flow with free surfaces. Despite the current great popularity of this method, every variant of the SPH does not meet all requirements on spatial discretization schemes.

Chapter 7 provides a brief introduction to Peridynamics according to Silling (2000) and Silling et al. (2007). This scheme is a rather new meshfree method based on computational approaches from Molecular Dynamics. After a short description, the fulfillment of the requirements on spatial discretization schemes is checked. Additionally, it is shown that Peridynamics corresponds to a Galerkin method.

Chapter 8 deals with the numerical simulation of Selective Laser Melting. First, the most important physical phenomena and the most important experimental observations are presented. The differential equations are discretized using SPH and the OTM method. Their effects on the results are discussed in detail. Another focus is the modeling of the laser together with a comparison of two different approaches.

Chapter 9 includes the description and the simulation of Direct Poly Printing according to Stieghorst (2017). First, the physical and chemical influencing factors, as well as the experimental results, are presented. After modeling the printing process, the equations are discretized using Peridynamics. Additionally, some results of the numerical simulation are presented.

References

C. Caviezel, R. Grünwald, S. Ehrenberg-Silies, S. Kind, T. Jetzke, M. Bovenschulte, *Additive Fertigungsverfahren (3-D-Druck)* (Vom Büro für Technikfolgen-Abschätzung beim Deutschen Bundestag, Arbeitsbericht, 2017)

R.A. Gingold, J.J. Monaghan, Smoothed Particle Hydrodynamics: theory and application to non-spherical stars. Mon. Not. R. Astron. Soc. **181**, 375–389 (1977)

C.W. Hirt, B.D. Nichols, Volume of fluid (VOF) method for the dynamics of free boundaries. J. Comput. Phys. **39**(1), 201–225 (1981)

C.W. Hull, Apparatus for production of three-dimensional objects by Stereolithography (1986). US Patent 4,575,330

Inc Wohlers Associates, *Wohlers Report 2017: Additive Manufacturing and 3D Printing State of the Industry* (Wohlers Associates, Fort Collins, 2017)

W.E. King, A.T. Anderson, R.M. Ferencz, N.E. Hodge, C. Kamath, S.A. Khairallah, A.M. Rubenchiks, Laser powder bed fusion Additive Manufacturing of metals; Physics, computational, and materials challenges. Appl. Phys. Rev. **2**(4), 041304 (2015)

C. Körner, A. Bauereiß, E. Attar, Fundamental consolidation mechanisms during Selective Beam Melting of powders. Modelling Simul. Mater. Sci. Eng. **21**(8), 085011 (2013)

S. Li, W.K. Liu, *Meshfree Particle Methods* (Springer, Berlin, 2007)

B. Li, F. Habbal, M. Ortiz, Optimal transportation meshfree approximation schemes for fluid and plastic flows. Int. J. Numer. Meth. Eng. **83**, 1541–1579 (2010)

M. Liu, G.R. Liu, *Particle Methods for Multi-scale and Multi-physics* (World Scientific, Singapore, 2015)

L.B. Lucy, Numerical approach to testing the fission hypothesis. Astron. J. **82**, 1013–1024 (1977)

M.A. Russell, A. Souto-Iglesias, T.I. Zohdi, Numerical simulation of laser fusion additive manufacturing processes using the SPH method. Comput. Methods Appl. Mech. Eng. **341**, 163–187 (2018)

B. Schoinochoritis, D. Chantzis, K. Salonitis, Simulation of metallic powder bed additive manufacturing processes with the finite element method: a critical review. Proc. Inst. Mech. Eng. Part B: J. Eng. Manuf. **231**(1), 96–117 (2017)

S.A. Silling, Reformulation of elasticity theory for discontinuities and long-range forces. J. Mech. Phys. Solids **48**(1), 175–209 (2000)

S.A. Silling, M. Epton, O. Weckner, J. Xu, E. Askari, Peridynamic states and constitutive modeling. J. Elast. **88**(2), 151–184 (2007)

J. Stieghorst, Entwicklung eines additiven Fertigungsverfahrens für Silikonkautschuke zur Herstellung von flexiblen Neuroimplantaten in der Medizintechnik. Ph.D. thesis, Leibniz Universität Hannover, Germany (2017)

C.L. Ventola, Medical applications for 3D printing: current and projected uses. Pharm. Ther. **39**(10), 704–711 (2014)

Z. Wang, W. Yan, W.K. Liu, M. Liu, Powder-scale multi-physics modeling of multi-layer multi-track selective laser melting with sharp interface capturing method. Comput. Mech. **63**(4), 649–661 (2019)

I. Yadroitsev. *Selective Laser Melting: Direct manufacturing of 3D-objects by Selective Laser Melting of Metal Powders* (Lambert Academic Publishing, Sunnyvale, 2009)

W. Yan, W. Ge, Y. Qian, S. Lin, B. Zhoun, W.K. Liu, F. Lin, G.J. Wagner, Multi-physics modeling of single/multiple-track defect mechanisms in electron beam selective melting. Acta Mater. **134**, 324–333 (2017)

W. Yan, S. Lin, O.L. Kafka, Y. Lian, C. Yu, Z. Liu, J. Yan, S. Wolff, H. Wu, E. Ndip-Agbor, M. Mozaffar, K. Ehmann, J. Cao, G.J. Wagner, W.K. Liu, Data-driven multi-scale multi-physics models to derive process-structure-property relationships for additive manufacturing. Comput. Mech. **61**(5), 521–541 (2018)

Chapter 2
Additive Manufacturing Processes

Additive Manufacturing (AM) refers to the fabrication of products layer-by-layer from formless raw materials based on a data triple (Fig. 2.1). This data triple consists of a 3D model of the part to be manufactured, the material recipe, and the process parameters. Additive Manufacturing is assigned a disruptive character since it is not based on knowledge of conventional manufacturing processes and hence establishes a new production paradigm (Acatech 2016). AM is often referred to as 3D printing. However, both terms have different meanings. Additive Manufacturing refers to the printing of functional components in an industrial environment, whereas 3D printing hallmarks the production of decorative objects in the customer-oriented segment. Here, both terms are used synonymously. Besides the description of different printing technologies and their applications, this chapter shows the potentials, as well as the gaps and needs of this still very young technology.

2.1 Potentials

In Additive Manufacturing, the materials are available in small dimensions. The part is produced by joining these small units together. Hence, significantly more design options in terms of geometry and material properties are available. The production of functional gradient materials can be simplified considerably. Also, the weight of a structure can be reduced leading to less material waste and energy consumption Huang et al. (2015).

AM contributes to the digitalization of manufacturing. The production depends mainly on the data triple which can be sold online or streamed. The data triple can be created locally and shared via cloud from different developers at different locations.

C. Weißenfels, *Simulation of Additive Manufacturing using Meshfree Methods*,
Lecture Notes in Applied and Computational Mechanics 97,
https://doi.org/10.1007/978-3-030-87337-0_2

Fig. 2.1 Additive Manufacturing: Direct fabrication of products from a CAD model based on a data triple

The production is only based on a 3D printing machine. The use of molds or assembly work is no longer necessary. The machine can be placed anywhere and the manufacturing process can be controlled online from another location. Alternatively, the data triples can be stored in a cloud and released to any user.

Overproduction can be avoided, as Additive Manufacturing enables the production on demand. Storing spare parts is therefore superfluous and stock holding can be reduced. Many products no longer have to be transported unnecessarily, as they can be manufactured directly on site. Since even complex geometries can be fabricated as a single piece, the number of individual parts is reduced compared to conventional processes (Caviezel et al. 2017).

Another big potential of Additive Manufacturing is the cost-effective fabrication of small lot sizes. The number of products has only a small influence on the manufacturing costs since only the data triple has to be adapted. Hence, mass customization, i.e. the fabrication of individual products for customers at the price of a product from mass production, can be realized with AM. In addition, the time to market for products is reduced. Designs at different development stages can be fabricated and tested directly. Thus, several iterations can be realized within a very short time. When the product has its final form, it can be offered directly on the market (Tofail et al. 2018).

2.2 Printing Technologies

All 3D printing technologies have the same data preparation, which can be divided into eight steps according to (Acatech 2016). First, the part is designed with a CAD program. Afterward, the data is converted into the .stl format with the surface approximated by triangles. Errors in the conversion must still be corrected manually. In the next step, the object is placed virtually in the build envelope. In the case of overhangs, support structures must be established for some methods to prevent undesired deformations and residual stress. This step requires a deep understanding of the process. Slicing cuts the body into parallel planes. For each layer, the exact sequence of print head movement must be defined so that each layer can be filled with material. This step is called hatching. Finally, the data triple can be generated.

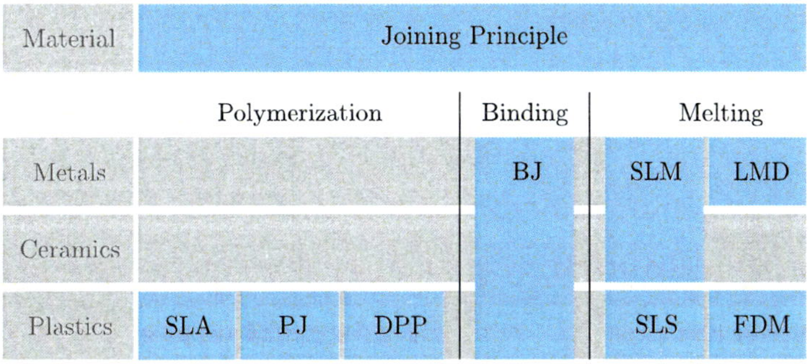

Fig. 2.2 Classification of printing technologies in Additive Manufacturing. SLA: Stereolithoraphy, PJ: Polymer Jetting, DPP, Direct Poly Printing, BJ: Binder Jetting, SLM: Selective Laser Melting, SLS: Selective Laser Sintering, EBM: Electron Beam Melting, FDM: Fused Deposit Modeling

3D printers can be classified according to e.g. the joining principle (Fig. 2.2). The selection of the printing process depends on the required build envelope, the fabrication speed, the possible layer heights, the resolution of the material, the printable materials, the accuracy of the process, and the costs (Huang et al. 2015). The focus lies here in the modeling of Selective Laser Melting and Direct Poly Printing. Therefore, the joining principles of melting and photopolymerization are presented in more detail.

2.2.1 Melting

The printing processes based on melting can be divided into three different categories:

- Material extrusion
- Powder bed fusion
- Material deposition

The best-known material extrusion technology is Fused Deposition Modeling (FDM) which is also called Fused Layer Manufacturing (FLM). The wire-shaped material is heated in the print head and deposited on the substrate or the previous layer via a nozzle (Fig. 2.3). This 3D printing process is often used in home applications. The materials that can be processed are limited to plastics. Mostly, amorphous thermoplastics such as acrylonitrile butadiene styrene (ABS), polylactic acid (PLA), or polyimide (PI) come into use (Acatech 2016).

In Selective Laser Sintering (SLS) and Selective Laser Melting (SLM), the material is in the form of a powder that is evenly distributed in a bed before processing (Fig. 2.3). The height of the powder layer is an additional process parameter. In Selective Laser Sintering, the bed consists of materials with high and low melting temperatures. The laser melts the latter. The movement of the unmelted particles causes the compaction of the material (Gu et al. 2012). Due to the low tempera-

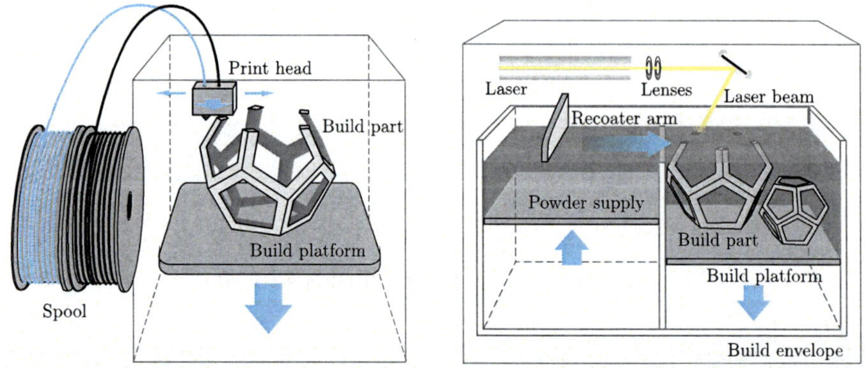

Fig. 2.3 Schematic illustration of (left) Fused Deposition Modeling and (right) Selective Laser Melting. (Drawings based on figures from www.additively.com)

ture, semi-crystalline thermoplastics such as polyamide (PA) or polyaryletherketone (PAEK) can be processed in addition to metallic multi-material components (MMC). The selection of the process parameters and of the material composition strongly influences the properties of the printed part. Mostly, a post-treatment is necessary for SLS.

Selective Laser Melting was developed to process pure weldable metals, to fabricate materials with a high density, and to avoid post-treatment (Gu et al. 2012). The formless material is treated in the same way as SLS, but now the material is completely melted. With this technology, parts possessing densities of over 99% can be achieved. Further details on this process can be found in Chap. 8.

Metallic components can also be produced using material deposition. Directed Light Fabrication (DLF) and Laser Engineered Net Shaping (LENS) (Lewis and Schlienger 2000) correspond to this type. The material is available either as a powder or as a wire and is fed directly to the laser beam. The resulting melt is deposited directly at the corresponding location. The print head can usually move in 5 axes and allows a 3D print of any geometry. Different weldable metals and metal alloys can be processed.

2.2.2 *Photopolymerization*

Photopolymerization is limited to the 3D printing of polymers. The aggregate state of the materials is always liquid. In contrast to melting, the heat source accelerates a chemical reaction that hardens the material in a targeted manner. Various printing concepts are available for this purpose:

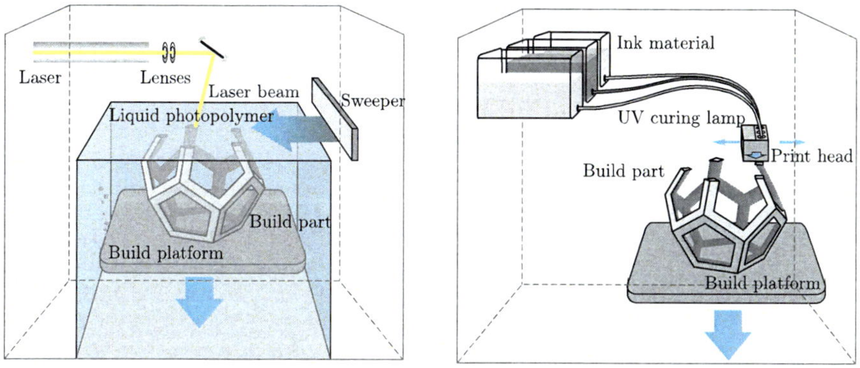

Fig. 2.4 Schematic illustration of (left) Stereolithography and (right) Polyjet Modeling. (Drawings based on figures from www.additively.com)

- VAT Photopolymerization
- Material jetting
- Material extrusion

Stereolithography (SLA) is a prominent representative of VAT photopolymerization. The liquid covers most of the build envelope. A controllable UV laser hardens parts of the liquid layer (Fig. 2.4). The height of a layer is adjusted via a moving platform. The materials used in this process have so far been limited to thermosets such as acrylic or epoxy resins.

Alternatively, these materials can be printed using Polyjet Modeling (PJ). An ink jet head deposits the photopolymer drop by drop. A movable light source ensures targeted curing (Fig. 2.4). Different materials can be easily mixed using different nozzles. Waxes are often applied to create the desired cavities in the part during the printing process.

Biocompatible silicone rubbers are often used for neural implants and cannot be accurately printed with conventional methods so far. For this reason, Stieghorst (2017) has developed a special deposition process that is here referred to as Direct Poly Printing (DPP). The material is deposited in viscous form onto the substrate or the already printed layers via a dispenser attached to a movable robot arm. A laser source accelerates the cross-linking of the silicone. More information about this 3D printing process can be found in Chap. 9.

2.3 Applications

The first 3D printed components were only used for the visualization of objects. Prototyping is still the most common application of AM today (Caviezel et al. 2017). In addition, Additive Manufacturing has established itself in the fabrication of spare parts or special tools that are not so frequently required, since the production through

3D printing is more profitable. In these contexts, the terminologies rapid prototyping, rapid repair, or rapid tooling are often used. The addition rapid is due to the elimination of time-consuming work steps, such as the assembly of single parts or the production of casting molds, which are not required in Additive Manufacturing (Acatech 2016). However, nowadays, Additive Manufacturing has also established itself as a production technology for mainly small lot sizes.

2.3.1 Aerospace Industry

In the aerospace industry, the requirements on the material are very high due to the temporarily very high ambient temperatures. The weight must be reduced to a minimum to ensure low consumption operation. On the other hand, the lot size is very small. Additive Manufacturing is an ideal production technology for optimized components, which often have complex geometries. For example, the exhaust gas ventilation of turbines in helicopters or turbine blades with integrated ventilation systems is usually additively manufactured (Guo and Leu 2013).

Remark With an aircraft lifetime of 30 years, every kilogram of weight saved reduces kerosene consumption by 3900–5800 L. This is equivalent to a reduction of carbon dioxide emissions by 11–16 tons (Huang et al. 2013).

2.3.2 Automotive Industry

Due to the current economic fabrication, the series production of ordinary car parts using Additive Manufacturing is not conceivable in the foreseeable future. However, this technology is occasionally used. For example, new, high-performance heat exchangers possessing complex geometries are additively manufactured in series production. A 3D print of these components is cheaper compared to alternative processes (Caviezel et al. 2017). 3D printed components are mostly used as special tools or assembly aids in automotive engineering. Due to the time savings in product development, prototypes are now mainly manufactured additively. Similar to the aerospace industry, the lot size of luxury cars or motorsports vehicles is very low. Particularly in the latter, lightweight components with special requirements on geometry and material properties are required, which are usually produced by 3D printing machines (Guo and Leu 2013).

2.3.3 Biomedical Technology

Every human being has a special anatomy. For implants and prostheses to optimally replace the missing human function, they must be tailor-made for each patient. Additive Manufacturing offers the possibility of fabricating the individual geometries cost-effectively compared to conventional manufacturing processes.

A congenital tracheobronchomalacia was successfully treated with the aid of a 3D printed tracheal splint (Zopf et al. 2013). The infant's too thin bronchial tube was sutured to the implant to correct the deformity. The elasticity of the splint allowed the natural growth of the infant. After 21 days, the bronchial tube was fully functional. One year after the surgery, no more negative abnormalities could be detected.

In the case of malpositions in the bone, scaffolds are often inserted to stimulate bone growth. At first, these scaffolds take over the bearing capacity of the bone. The scaffold have to be biocompatible and must optimally stimulate cell growth (Wang et al. 2007; Bose et al. 2012). A study of cell growth using different 3D printed scaffolds from hydroxyapatites can be found in Leukers et al. (2005).

In dental and hearing aid technology, Additive Manufacturing has been established for a long time (Caviezel et al. 2017). Especially patient-specific crowns and bridges can nowadays be fabricated economically in large quantities. The same applies to high-quality, individualized ear molds of a hearing aid. Another area of application for Additive Manufacturing in medical technology is the production of special clinical tools for surgeries.

2.3.4 Bioprinting

Bioprinting describes the fabrication of living organs and tissues. A successful 3D print eliminates the need for transplantation. The human tissue consists of cells with an extracellular matrix. The challenge of 3D printing is the reproducibility of the complex heterogeneous microstructure of the human tissue (Murphy and Atala 2014). Particular attention must be paid to the formation of the vascular system to ensure oxygen and nutrient supply to the organs (Visconti et al. 2010). For the accurate reproduction of the extracellular matrix, a suitable substitute material must be chosen. This consists mostly of natural or synthetic polymers. During bioprinting, very small cell units are deposited, which grow together independently. In addition to the exact reconstruction, the ability of the tissue to self-assemble must be supported by the printed cells (Murphy and Atala 2014).

For bioprinting, conventional technologies are adjusted. Ink jetting methods are often reconstructed to deposit spherical droplets consisting of extracellular matrix and stem cells. Alternatively, extrusion-based or light-supporting printing methods are used (Zhu et al. 2016).

Printing of complete organs that can be transplanted into the human organism is still a long way off (Ozbolat 2015). The application currently extends more to the simulation of small human structures to test new clinical procedures or treatments. These tests are designed to better understand human reaction behavior. Thus, animal experiments can be reduced or avoided. First successful attempts to print a complete 3-dimensional vascularized tissue on a small scale are documented in Miller et al. (2012) and Kolesky et al. (2014). In the former, a combined manufacturing process of 3D printing and casting was developed. In Kolesky et al. (2014), the tissue was fabricated completely by 3D printing.

Remark According to Ventola (2014), 154,324 patients in the USA were waiting for a donor organ in 2009. Only 27,996 patients underwent transplantation. 8,863 people died while being on the waiting list.

2.3.5 Defense Industry

Similar to the aerospace industry, the defense industry mainly requires systems with small lot sizes. In addition, individualized equipment, especially for protection, is needed. Some products are only desired on demand and directly on site. Additive Manufacturing thus proves to be an ideal production process in the defense industry. 3D printed parts are now said to be used in missiles, drones, and military engines (Fey 2016).

Additive Manufacturing makes it easy to bypass government protection and control systems. Weapon systems can be fabricated uncontrolled in secrecy, or like e.g. firearms made of plastic can pass control systems undetected (Caviezel et al. 2017). In addition, there is the problem of dual-use. For example, additively manufactured drones or engines can come into operation not only for civilian but also for military applications. Therefore, there are considerations to make high-performance additive production machines subject to approval (Caviezel et al. 2017).

2.3.6 Civil Engineering

Every building is unique. Compared to previous applications, the build envelope is significantly larger. The use of Additive Manufacturing in civil engineering is primarily aimed at the 3D printing of concrete structures (Bos et al. 2016). With this method, not only the shape but also the properties can be optimized. The outer layer can consist of self-cleaning concrete with resistance against aging. The inner structure guarantees the bearing capacity and enables good insulation properties. On the inside, the concrete can offer a good acoustic. 3D printing can also reduce the amount of concrete. Since the production of cement is very energy-intensive, large savings can be achieved.

Additive Manufacturing of building parts does not require formwork. On the other hand, this advantage is one of the major challenges in the development of

3D printers. Free printing means that the hydration of concrete must be controlled to guarantee the planned width and good surface quality of individual layers. In conventional fabrication, a vibrator ensures compaction. This must also be guaranteed if the structure is printed.

2.4 Gaps and Needs

One of the biggest barriers to Additive Manufacturing is the lack of reproducibility (Acatech 2016). The geometry and material properties are often different, although the fabrication is based on the same data triple. The properties of the printed part often do not correspond to the planned specifications. In addition, the slow fabrication speed and the lack of economic efficiency hinder the breakthrough of this technology as a powerful manufacturing process. In most cases, post-treatment is also required to improve surface quality or to correct unwanted deformations.

Additive Manufacturing is an interplay of several disciplines (Fig. 2.5), in which several challenges have to be solved.

2.4.1 Machines and Materials

Additive Manufacturing systems are mostly based on rapid prototyping machine architectures (Huang et al. 2015). However, the requirements of industrial manufacturing are much higher. Besides ensuring a high quality and fabrication speed, new 3D printers have to work autonomously. Particularly with powder bed processes, contact with the very fine metal powder must be avoided so that no particles enter the human organism. The post-processing must be integrated to keep manual intervention to a minimum. The maintenance intervals for 3D printers are still very short which also increases the costs. For example, filters have to be changed frequently. The intervals for new printers must be extended significantly (Acatech 2016).

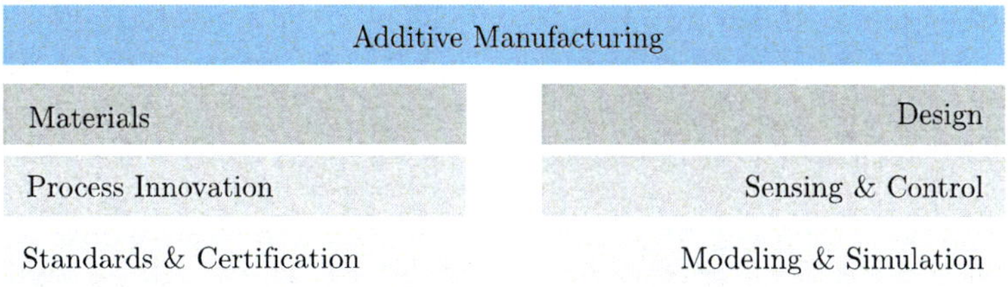

Fig. 2.5 Additive Manufacturing as an interplay of several disciplines

Additive Manufacturing makes it possible to separate design (data triple) from production. To enable smaller companies an access to this market, suitable platforms such as clouds are needed to exchange data or print jobs from other companies.

Mostly, standard materials are processed in Additive Manufacturing. However, the requirements for the material have increased. The selection of materials for 3D printing is still limited. It is often very difficult to produce materials, such as β-titanium, in the desired formless structure (Huang et al. 2015). New materials that can be processed by 3D printers must be developed. Additionally, in the production of components, it is necessary to know the interactions between the materials during melting.

Another barrier is the lack of standards. For example, the characteristics or the process parameters for materials are different for machines from different manufacturers. The introduction of standards, the certification of machines and materials is, in addition to the further development of printers and materials, another necessary step for the acceptance and breakthrough of Additive Manufacturing in the industry.

2.4.2 Design and Control

The planning of a part and the production process is carried out purely on a computer. The individual components are designed using a CAD program. During planning, the relationships between the process, the structure, and the material properties must be taken into account. This requires user-friendly software that supports the engineer in the design of new parts and automatically determines the printing process and its parameters.

To understand the relationship between process, structure, and properties, all effects, which occur during the printing process, must be understood. This requires the modeling of the entire process. The large temperature rates lead to strong changes in the microstructure. The models must be able to reproduce all these aspects. By comparing the simulation results with experimental data, the approaches can be validated. For this, a lot of experimental data need to be available. Suitable sensors must be installed in the build envelope to record temperature and displacements in the structure. If the simulations are accurate enough, not only can the printing process be optimized, but the process parameters can also be adapted in real-time during printing. Sensors characterize the current state and suitable parameters are determined based on simulation data. For this vision, efficient and highly accurate simulation methods must be available.

References

Acatech. *Additive Fertigung*. Acatech Position. Deutsche Akademie der Technikwissenschaften, Nationale Akademie der Wissenschaften Leopoldina, Union der deutschen Akademien der Wissenschaften (2016)

F. Bos, R. Wolfs, Z. Ahmed, T. Salet, Additive manufacturing of concrete in construction: potentials and challenges of 3d concrete printing. Virtual Phys. Prototyp. **11**(3), 209–225 (2016)

S. Bose, M. Roy, A. Bandyopadhyay, Recent advances in bone tissue engineering scaffolds. Trends Biotechnol. **30**(10), 546–554 (2012)

C. Caviezel, R. Grünwald, S. Ehrenberg-Silies, S. Kind, T. Jetzke, M. Bovenschulte, *Additive Fertigungsverfahren (3-D-Druck)* (Vom Büro für Technikfolgen-Abschätzung beim Deutschen Bundestag, Arbeitsbericht, 2017)

M. Fey, *Waffen aus dem 3D-Drucker: Additives Fertigen als sicherheitspolitisches Risiko?*, vol. 9 (DEU, 2016)

D.D. Gu, W. Meiners, K. Wissenbach, R. Poprawe, Laser Additive Manufacturing of metallic components: materials, processes and mechanisms. Int. Mater. Rev. **57**(3), 133–164 (2012)

N. Guo, M.C. Leu, Additive Manufacturing: technology, applications and research needs. Front. Mech. Eng. **8**(3), 215–243 (2013)

Y. Huang, M.C. Leu, J. Mazumder, A. Donmez, Additive Manufacturing: current state, future potential, gaps and needs, and recommendations. J. Manuf. Sci. Eng. **137**(1), 014001 (2015)

S.H. Huang, P. Liu, A. Mokasdar, L. Hou, Additive Manufacturing and its societal impact: a literature review. Int. J. Adv. Manuf. Technol. **67**(5–8), 1191–1203 (2013)

D.B. Kolesky, R.L. Truby, A.S. Gladman, T.A. Busbee, K.A. Homan, J.A. Lewis, 3D bioprinting of vascularized, heterogeneous cell-laden tissue constructs. Adv. Mater. **26**(19), 3124–3130 (2014)

B. Leukers, H. Gülkan, S.H. Irsen, S. Milz, C. Tille, M. Schieker, H. Seitz, Hydroxyapatite scaffolds for bone tissue engineering made by 3D printing. J. Mater. Sci. - Mater. Med. **16**(12), 1121–1124 (2005)

G.K. Lewis, E. Schlienger, Practical considerations and capabilities for laser assisted direct metal deposition. Mater. & Design **21**(4), 417–423 (2000)

J.S. Miller, K.R. Stevens, M.T. Yang, B.M. Baker, D.T. Nguyen, D.M. Cohen, E. Toro, A.A. Chen, P.A. Galie, X. Yu et al., Rapid casting of patterned vascular networks for perfusable engineered three-dimensional tissues. Nat. Mater. **11**(9), 768 (2012)

S.V. Murphy, A. Atala, 3D bioprinting of tissues and organs. Nat. Biotechnol. **32**(8), 773 (2014)

I.T. Ozbolat, Bioprinting scale-up tissue and organ constructs for transplantation. Trends Biotechnol. **33**(7), 395–400 (2015)

J. Stieghorst, Entwicklung eines additiven Fertigungsverfahrens für Silikonkautschuke zur Herstellung von flexiblen Neuroimplantaten in der Medizintechnik. Ph.D. thesis, Leibniz Universität Hannover, Germany (2017)

S.A.M. Tofail, E.P. Koumoulos, A. Bandyopadhyay, S. Bose, L. O'Donoghue, C. Charitidis, Additive Manufacturing: Scientific and technological challenges, market uptake and opportunities. Mater. Today **21**(1), 22–37 (2018)

C.L. Ventola, Medical applications for 3D printing: current and projected uses. Pharm. Therap. **39**(10), 704–711 (2014)

R.P. Visconti, V. Kasyanov, C. Gentile, J. Zhang, R.R. Markwald, V. Mironov, Towards organ printing: engineering an intra-organ branched vascular tree. Expert Opin. Biol. Ther. **10**(3), 409–420 (2010)

H. Wang, Y. Li, Y. Zuo, J. Li, S. Ma, L. Cheng, Biocompatibility and osteogenesis of biomimetic nano-hydroxyapatite/polyamide composite scaffolds for bone tissue engineering. Biomaterials **28**(22), 3338–3348 (2007)

W. Zhu, X. Ma, M. Gou, D. Mei, K. Zhang, S. Chen, 3D printing of functional biomaterials for tissue engineering. Curr. Opin. Biotechnol. **40**, 103–112 (2016)

D.A. Zopf, S.J. Hollister, M.E. Nelson, R.G. Ohye, G.E. Green, Bioresorbable airway splint created with a three-dimensional printer. N. Engl. J. Med. **368**(21), 2043–2045 (2013)

Chapter 3
Differential Equations

Additive Manufacturing involves the interaction of different physical processes. In some printing technologies, an additional chemical reaction takes place. A thermo-mechanical process with phase change is present if the 3D-print is based on melting. In the case of photopolymerization, an external heat source accelerates the hardening of a liquid due to a chemical reaction. Hence, mechanical, thermal, and chemical fields need to be coupled.

This chapter summarizes the differential equations, which are required for the modeling in the field of Additive Manufacturing. Since each solution scheme is based either on the strong and or the weak form of the differential equation, both formulations are presented. Only the most important equations are summarized. A detailed introduction into continuum mechanics can be found e.g. in Truesdell and Noll (2004), Truesdell and Toupin (1960), Silhavy (1997), Malvern (1969) or Holzapfel (2000). Monographs especially addressing material modeling are Haupt (2002) or Krawietz (1986). A more mathematical introduction into elasticity presents Marsden and Hughes (1983).

In Additive Manufacturing, boundary layer effects play a decisive role. These include, for example, surface tension, Marangoni convection, and the transition between the individual aggregate states. Hence, the mathematical formulations of boundary effects are introduced as well.

3.1 Modeling of Continua

The modeling of continua is mainly based on a unique decomposition of a body into rigid points. Each point P of the body $\mathcal{B}$ possesses a position vector $\mathbf{x}$ in the Euclidean space $\mathbb{E}^3$ and a mass. Due to loading, the body deforms and the individual mass

C. Weißenfels, *Simulation of Additive Manufacturing using Meshfree Methods*, Lecture Notes in Applied and Computational Mechanics 97,
https://doi.org/10.1007/978-3-030-87337-0_3

points move. The sum of all position vectors is called configuration. Mathematically, a configuration is a bijective mapping $\boldsymbol{\kappa}$ of local points into the Euclidean space

$$\begin{aligned} \boldsymbol{\kappa} : \mathcal{B} \rightarrow \mathbb{E}^3, \qquad \mathbf{x} = \boldsymbol{\kappa}(P, t) \in \mathbb{E}^3 \\ B_t = \boldsymbol{\kappa}(\mathcal{B}, t) = \{\boldsymbol{\kappa}(P, t),\, P \in B,\, t \in \mathbb{R}\} \subset \mathbb{E}^3. \end{aligned} \tag{3.1}$$

At any time t a body has a different configuration, which is abbreviated as B_t. The sum of all configurations from $t = t_0$ to the current time is called motion. Two configurations are of special interest. The sum of all position vectors at the time $t = t_0$ indicates the initial configuration B_0. The current configuration B_t represents the body at the actual time t. Based on (3.1) a mapping φ of position vectors from the initial to the current configuration can be defined

$$\varphi : \mathbb{E}^3 \rightarrow \mathbb{E}^3, \quad \mathbf{x} = \varphi(\mathbf{X}, t) = \boldsymbol{\kappa}(\boldsymbol{\kappa}^{-1}(P, 0), t) \in \mathbb{E}^3. \tag{3.2}$$

In continuum mechanics, two different descriptions are available. Either the differential equations are formulated with respect to a fixed material point (Lagrangian description, see Fig. 3.1) or a fixed point in space (Eulerian description). In the latter, the material time derivative has an additional convective part which can lead to numerical problems (Brooks and Hughes 1982). If mesh-based methods are employed to simulate large deformations, which occur e.g. with fluid flows, the differential equations are mostly formulated with respect to a fixed point in space. Employing meshfree methods the equations for fluid flow can also be solved using the Lagrangian description. This is especially advantageous if free surfaces are present like in the case of Additive Manufacturing. For this reason, the equations, which are presented here, are based solely on this description.

The differential equations can be formulated for the whole continuum or locally for each mass point. In addition, a weak form also exists. Since meshfree methods can be based on either the weak or the strong form, both descriptions are given.

Remark Besides the classical description of continua employing a decomposition of the body into rigid mass points, some theories allow rotations of the particles. This form is called micro-polar or Cosserat theory. An extension to deformable particles

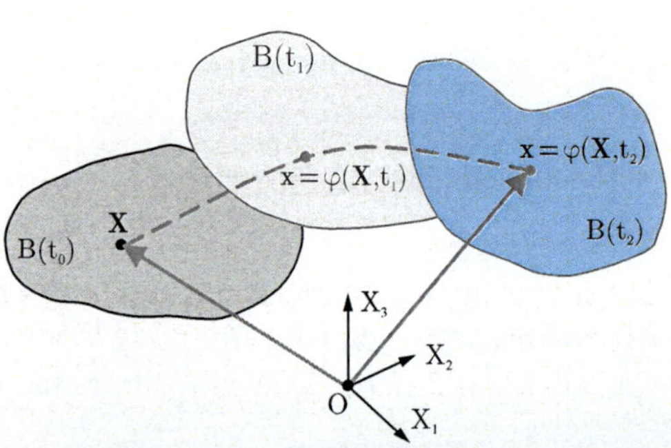

Fig. 3.1 Lagrangian description: Formulation of the differential equation with respect to mass points

describes a micromorphic formulation. An introduction to these theories can be found in Eringen and Kafadar (1976).

3.2 Mechanical Differential Equation

The solution function for describing the response behavior under mechanical loading is the displacement $\mathbf{u}$. This quantity can be specified for each mass point and corresponds to the difference between the current and the initial position vector, i.e. $\mathbf{u} = \mathbf{x} - \mathbf{X}$. The mechanical differential equation for calculating the displacements results from the balance of momentum together with proper constitutive and kinematic relations.

In this short overview, all equations are given in the local form and with respect to the current configuration. Alternative representations can be found in the aforementioned textbooks.

3.2.1 Strong Form for Compressible Materials

The balance of momentum postulates that the change of the momentum in time is equal to the sum of external forces. This statement requires the inertia term $\rho\ddot{\mathbf{u}}$ to be equal to the divergence of the Cauchy stress tensor $\boldsymbol{\sigma}$ and the specific dead load $\rho\bar{\mathbf{b}}$

$$\rho\ddot{\mathbf{u}} = \operatorname{div}\boldsymbol{\sigma} + \rho\bar{\mathbf{b}}. \tag{3.3}$$

The second term on the right hand side is calculated from the density ρ and the gravitational acceleration represented as the vector $\bar{\mathbf{b}}$. In order to fulfill the balance of angular momentum the Cauchy stress tensor must be symmetric. According to Walter Noll's definition for simple and homogeneous materials, the stress tensor is a function of the deformation gradient $\mathbf{F}$, the temperature Θ, and the temperature gradient (Truesdell and Noll 2004). Taking into account the 2nd law of thermodynamics, this dependence is reduced to $\mathbf{F}$ and Θ. In the case of chemical reactions, the description can be extended to the degree of cure α. For isotropic materials, the stress tensor can be defined as a function of the right Cauchy Green tensor $\mathbf{b}$. In case of hyper-elastic materials, the stress tensor is calculated from a potential defined as the specific strain energy function ψ

$$\boldsymbol{\sigma} = \rho\frac{\partial\psi\left(\mathbf{F}, \Theta, \alpha\right)}{\partial\mathbf{F}}\mathbf{F}^{\mathrm{T}} = 2\rho\mathbf{b}\frac{\partial\psi\left(\mathbf{b}, \Theta, \alpha\right)}{\partial\mathbf{b}}, \quad \mathbf{b} = \mathbf{F}\mathbf{F}^{\mathrm{T}}. \tag{3.4}$$

The term specific indicates a quantity that is related to the mass. Thus, this information is independent of the size of the system (Lüdecke and Lüdecke 2013). The constitutive equations depend on the materials used and the environmental conditions. Suitable models for ψ can be found in the sections on Additive Manufacturing in Chaps. 8 and 9. The deformation gradient provides the relationship to the solution function

$$\mathbf{F} = \frac{\partial \mathbf{x}}{\partial \mathbf{X}} = \operatorname{Grad}\mathbf{x} = \mathbf{1} + \operatorname{Grad}\mathbf{u}. \tag{3.5}$$

Assuming that the mass of the system remains constant during the process, the current density can be calculated from its initial value ρ_0 using the Jacobi determinant J

$$\rho = \frac{\rho_0}{J}, \qquad J = \det \mathbf{F}. \tag{3.6}$$

The derivative with respect to time yields to an alternative expression known as the continuity equation

$$\frac{\mathrm{d}}{\mathrm{d}t}(\rho J) = 0, \quad \text{or} \quad \dot{\rho} + \rho \operatorname{div}\mathbf{v} = 0. \tag{3.7}$$

Therein, the vector $\mathbf{v}$ denotes the velocity of the mass point.

3.2.2 *Strong Form for Incompressible Materials*

To characterize the material behavior, the kinematic and constitutive quantities can be divided into a volumetric, i.e. volume changing and an isochoric, i.e. volume preserving part. The decomposition of the deformation gradient is multiplicative, whereby the volumetric part depends only on the Jacobi determinant

$$\mathbf{F} := \mathbf{F}^{vol}\mathbf{F}^{iso}, \quad \text{where} \quad \mathbf{F}^{vol} = J^{\frac{1}{3}}\mathbf{1}, \quad \text{and} \quad \det \mathbf{F}^{iso} = 1. \tag{3.8}$$

The specific strain energy function can be divided additively into a volumetric and an isochoric term

$$\psi := \psi^{vol}(J) + \psi^{iso}\left(\mathbf{b}^{iso}\right), \quad \text{where} \quad \mathbf{b}^{iso} = J^{-\frac{2}{3}}\mathbf{b}. \tag{3.9}$$

The derivative with respect to the kinematical quantity results in a split of the stress tensor. The volumetric part corresponds to the hydrostatic pressure p. Compressive stresses are always negative in continuum mechanics. Since p is usually defined positively at pressure, a minus sign is introduced

$$\boldsymbol{\sigma} = -p\mathbf{1} + 2\rho\mathbf{b}\frac{\partial \psi^{iso}}{\partial \mathbf{b}}, \quad p = -\rho\frac{\partial \psi^{vol}}{\partial J}. \tag{3.10}$$

If the total volume does not change during loading at constant temperature, incompressible material behavior is present. This requirement constraints the density to be constant. Equations (3.6) and (3.7) simplify to

$$J - 1 = 0 \quad \text{or} \quad \operatorname{div} \mathbf{v} = 0. \tag{3.11}$$

In this case, the volumetric part of the strain energy function is zero. The hydrostatic pressure is calculated from the fulfillment of the incompressibility condition (3.11). This constraint can also be introduced into the specific strain energy function via a Lagrange multiplier λ, where λ corresponds to the hydrostatic pressure (e.g. Holzapfel 2000)

$$\psi := \psi(\mathbf{b}) - \lambda(J-1), \quad \text{where} \quad \mathbf{b} = \mathbf{b}^{iso}, \quad \text{and} \quad \lambda = p. \tag{3.12}$$

From (3.11) it follows directly that the right Cauchy Green tensor is isochoric. The derivative of the strain energy function with respect to the kinematic quantities leads to (3.10). The derivative with respect to pressure gives the incompressibility constraint.

Near a structure fluids show a viscous behavior. With the introduction of dissipative stresses (Coleman and Gurtin 1967), incompressible Newtonian fluids can be described as a function of the dynamic viscosity η (Malvern 1969)

$$\boldsymbol{\sigma} = -p\mathbf{1} + 2\eta\mathbf{d}, \qquad \mathbf{d} = \operatorname{grad}^{sym}\mathbf{v} = \frac{1}{2}\left(\operatorname{grad}\mathbf{v} + \operatorname{grad}^{\mathrm{T}}\mathbf{v}\right). \tag{3.13}$$

The viscous part is determined by the symmetric spatial velocity gradient $\mathbf{d}$. For incompressible fluids mostly the velocity $\mathbf{v}$ and the pressure p are chosen as the solution functions. In this case, the pressure is calculated by fulfilling the incompressibility condition $\operatorname{div}\mathbf{v} = 0$. For Newtonian fluids, the differential equation can be directly formulated in terms of the solutions functions

$$\rho\dot{\mathbf{v}} = -\operatorname{grad} p + 2\eta \operatorname{div} \operatorname{grad}^{sym}\mathbf{v} + \rho\bar{\mathbf{b}}. \tag{3.14}$$

3.2.3 *Boundary and Initial Conditions*

The supports and the loadings correspond to the mechanical boundary conditions. A distinction is made between the essential, i.e. the Dirichlet, and the natural, i.e. the Neumann boundary conditions. For the former, specifications are given for the solution function. For the latter, values for its derivative have to be known which is usually formulated in terms of an external stress vector $\bar{\mathbf{t}}$

$$\mathbf{u} = \bar{\mathbf{u}} \quad \text{on} \quad \partial B_t^{D_u}, \quad \boldsymbol{\sigma}\cdot\mathbf{n} = \bar{\mathbf{t}} \quad \text{on} \quad \partial B_t^{N_u}. \tag{3.15}$$

The boundary is either of a Dirichlet or a Neumann type, i.e. $\partial B_t^{D_\mathbf{u}} \cup \partial B_t^{N_\mathbf{u}} = \partial B_t$ and $\partial B_t^{D_\mathbf{u}} \cap \partial B_t^{N_\mathbf{u}} = \emptyset$. The normalized normal vector $\mathbf{n}$ at the surface points outwards. In the case of fluids, conditions are given for to the velocity. For contact with solids, the Dirichlet boundary is divided into a normal and a tangential part. In the normal direction, it is required that a fluid cannot flow into the solid

$$\mathbf{v} \cdot \mathbf{n} = 0 \quad \text{on} \quad \partial B_t^{D_\mathbf{u}}. \tag{3.16}$$

In the tangential direction, either the fluid particles stick to the surface of the solid or they slide on the boundary. In the first case, the velocity of the fluid particles in contact is zero. If the second case occurs, a constitutive equation for sliding must be formulated

$$\text{Stick:} \;\; \mathbf{v} = 0 \;\; \text{Slip:} \;\; \mathbf{v} \cdot \mathbf{t}_\alpha \neq 0 \;\; \text{on} \;\; \partial B_t^{D_\mathbf{u}}. \tag{3.17}$$

The vectors $\mathbf{t}_\alpha$ correspond to the two base vectors at the boundary surface ($\alpha = 1, 2$). If the displacement is the solution function, the differential equation is of 2nd order in time. In this case, the values for the displacements and the velocities at each mass point at time $t = t_0$ are to be given

$$\mathbf{u}\,(t = t_0) = \mathbf{u}_0, \quad \text{and} \quad \mathbf{v}\,(t = t_0) = \mathbf{v}_0 \quad \text{in} \quad B_t. \tag{3.18}$$

If the velocity is selected as the solution function, only $\mathbf{v}\,(t = t_0)$ has to be specified.

3.2.4 *Interface Effects*

In Additive Manufacturing, the dimensions are very small. If the material is in liquid form, boundary layer effects such as surface tension or Marangoni convection must be taken into account. If the temperature exceeds a critical value, part of the material evaporates. In this case, the recoil pressure must also be considered. More details on these phenomena can be found in Sect. 8.1. Mechanically, there is a jump of the stress vector between the two states of aggregation

$$\mathbf{t}^1 = \boldsymbol{\sigma}^1\mathbf{n} \neq \mathbf{t}^2 = \boldsymbol{\sigma}^2\mathbf{n}. \tag{3.19}$$

The direction of the jump is defined by the normalized normal vector $\mathbf{n}$ at the boundary layer. In this work, only the liquid phase is modeled during the liquid to gas transition. Likewise, wetting is not included. Thus, in (3.19), only the fractions for the liquid phase are significant, leading to the Neumann boundary condition (3.15). The stress vector can be divided into a normal and a tangential component. The difference from (3.19) in the normal direction corresponds to the surface tension and the recoil pressure p^{rec}. The difference in the tangential direction describes the

Marangoni convection. Here, the derivative with respect to the surface coordinates $\mathbf{x}^s$ is needed

$$\boldsymbol{\sigma}\mathbf{n} = -p\mathbf{n} + \mathbf{t}_T = -\left(2\gamma\kappa + p^{rec}\right)\mathbf{n} + \frac{\partial\gamma}{\partial\mathbf{x}^s}. \tag{3.20}$$

The component due to surface tension follows from the Young Laplace equation. This pressure component is determined from the mean curvature κ and the coefficient of surface tension γ. This parameter is a function of temperature Θ, which in turn can be different at each position vector $\mathbf{x}_s$ at the surface, i.e. $\gamma := \gamma(\Theta(\mathbf{x}_s))$. A non-constant distribution of the temperature at the surface triggers the Marangoni convection. A detailed derivation of the resulting contributions can be found in Landau and Lifshitz (1987). The recoil pressure often results from an estimate using the Clausius-Clapeyron relation. The corresponding equation together with a reference to further literature can be found in Sect. 8.1.6.

3.2.5 *Weak Form*

Galerkin solution schemes are based on the weak form of the differential equation. Therefore, the equations are multiplied with a test function $\boldsymbol{\eta}$ and integrated over the volume of the body. This description also corresponds to the principle of weighted residuals (e.g. Schwarz 1991). Using Gauss's integral theorem, the weak form contains only first derivatives

$$G(\boldsymbol{\eta}, \mathbf{u}) = \int_{B_t} \left(\operatorname{grad}\boldsymbol{\eta}\cdot\boldsymbol{\sigma} - \rho\boldsymbol{\eta}\cdot\left(\bar{\mathbf{b}} - \ddot{\mathbf{u}}\right)\right)\,dv - \int_{\partial B_t^{N_\mathbf{u}}} \boldsymbol{\eta}\cdot\bar{\mathbf{t}}\,da = 0. \tag{3.21}$$

The Neumann boundary condition is directly included in the formulation. Since the order of the differential equation is reduced by one, only the Dirichlet boundary conditions need to be satisfied together with $\boldsymbol{\eta} = 0$ on $B_t^{D_\mathbf{u}}$. A detailed derivation of the weak form can be found in monographs on the Finite Element Method, such as Hughes (1987) or Wriggers (2008).

3.3 Thermal Differential Equation

Additive Manufacturing processes are based on the control of the temperature Θ in the body. The differential equation is derived from the 1st law of thermodynamics. In this section, only the most important equations are presented. An introduction to thermodynamics can be found e.g. in Lüdecke and Lüdecke (2013) or Baehr and Kabelac (2016) and a continuum mechanical treatment in Silhavy (1997), Truesdell and Noll (2004), Truesdell and Toupin (1960) or Miehe (1988). To be consistent with the previous section all equations are given in the local form and with respect to the current configuration.

3.3.1 Strong Form

The 1st law of thermodynamics postulates that the change of the total energy in time corresponds to the sum of mechanical work and external heat. With the introduction of the specific internal energy u the 1st law of thermodynamics can be reformulated. The internal stress power $\boldsymbol{\sigma} \cdot \mathbf{d}$, the change of the heat flow $\mathbf{q}$, and the internal heat source r reasons a change of u

$$\rho \dot{u} = \boldsymbol{\sigma} \cdot \mathbf{d} - \operatorname{div} \mathbf{q} + \rho r. \tag{3.22}$$

The specific inner energy can be related to the specific strain energy function by a Legendre transformation (e.g. Silhavy 1997)

$$\psi = u - \Theta \eta. \tag{3.23}$$

The quantity η corresponds to the specific entropy and can be described as a function of the deformation gradient and the temperature $\eta := f(\mathbf{F}, \Theta)$. Here, it has to be noted that Θ corresponds to the absolute temperature, i.e. $\Theta \geq 0$. From the 2nd law of thermodynamics the amount of internal dissipation D_{int} can be specified

$$D_{int} = \boldsymbol{\sigma} \cdot \mathbf{d} - \rho \dot{\psi} - \rho \dot{\Theta} \eta \geq 0. \tag{3.24}$$

Since D_{int} is equal to zero for elastic materials, the stress (3.4), and the specific entropy can be derived from the specific strain energy function

$$\eta(\mathbf{F}, \Theta) = -\frac{\partial \psi}{\partial \Theta}. \tag{3.25}$$

With these relations, the 1st law of thermodynamics can be rewritten

$$\rho \Theta \dot{\eta} = -\rho \Theta \frac{\partial^2 \psi}{\partial \Theta \partial \mathbf{F}} \cdot \dot{\mathbf{F}} - \rho \Theta \frac{\partial^2 \psi}{\partial^2 \Theta} = D_{int} - \operatorname{div} \mathbf{q} + \rho r. \tag{3.26}$$

A detailed derivation of the above equation can be found e.g. in Truesdell and Noll (2004). Also the heat flow $\mathbf{q}$ is a constitutive quantity. The relation to the solution function Θ is given by Fourier's law

$$\mathbf{q} = -\mathbf{K} \operatorname{grad} \Theta. \tag{3.27}$$

In case of isotropic material behavior, the 2nd order tensor **K** reduces to the thermal conductivity k multiplied by the unit tensor. The derivative of entropy with respect to temperature corresponds to the specific heat capacity

$$c_{\mathbf{F}} = \Theta \frac{\partial \eta}{\partial \Theta} = -\Theta \frac{\partial^2 \psi}{\partial^2 \Theta}. \tag{3.28}$$

It describes the amount of heat necessary to increase the temperature by 1 while keeping the deformation constant. The derivative of specific entropy with respect to the deformation gradient corresponds to the latent heat

$$\mathbf{L}_{\mathbf{F}} = \Theta \frac{\partial \eta}{\partial \mathbf{F}} \mathbf{F}^{\mathrm{T}} = -\Theta \frac{\partial^2 \psi}{\partial \Theta \partial \mathbf{F}} \mathbf{F}^{\mathrm{T}}. \tag{3.29}$$

The 2nd order tensor $\mathbf{L}_{\mathbf{F}}$ is symmetric due to the definition of the Cauchy stress tensor (Silhavy 1997). If isotropic material behavior is present, this term reduces to a scalar $L_{\mathbf{F}}$ multiplied by the unit tensor. The latent heat (3.29) describes the supply of heat necessary to increase a relative deformation by 1 at a constant temperature. For isotropic materials, this deformation is purely volumetric. This effect is observed in elastomers and is often referred to as the Gough Joule effect (Holzapfel 2000 or Miehe 1988). With the definition of the individual quantities, the 1st law of thermodynamics for isotropic materials can be specified more precisely

$$\rho L_{\mathbf{F}} \operatorname{tr} \mathbf{d} + \rho c_{\mathbf{F}} \dot{\Theta} = D_{int} - \operatorname{div} \mathbf{q} + \rho r. \tag{3.30}$$

The heat capacity and the latent heat can also be formulated with respect to stress. If heat conduction is prevented, the supply or removal of heat is only possible through the external source r. For elastic materials, the internal dissipation is equal to zero and it has to hold

$$r = c_{\mathbf{F}} \dot{\Theta} + L_{\mathbf{F}} \operatorname{tr} \mathbf{d} = c_{\sigma} \dot{\Theta} + L_{\sigma} \operatorname{tr} \dot{\boldsymbol{\sigma}}. \tag{3.31}$$

The quantity c_{σ} indicates the heat required to increase the temperature of a mass by 1 at constant stress. The latent heat L_{σ} describes the heat necessary to increase the volumetric stress by one at a constant temperature. A relationship between stress and deformation-dependent quantities can be found in Silhavy (1997) or Truesdell and Toupin (1960).

3.3.2 *Boundary and Initial Conditions*

The Dirichlet conditions correspond to prescribed values of the temperature at the boundary. Neumann boundary conditions are imposed by the heat flow over the surface. From a practical point of view, instead of the absolute temperature, usually only the temperature increase ϑ with respect to a spatially constant reference temperature Θ_0 is considered

$$\vartheta = \Theta - \Theta_0. \tag{3.32}$$

For all terms that appear as derivatives, the absolute temperature can be replaced by the relative contribution without changing the balance equations

$$\dot{\vartheta} = \dot{\Theta}, \qquad \operatorname{grad} \vartheta = \operatorname{grad} \Theta. \tag{3.33}$$

Since ϑ is usually chosen as the solution function, the boundary conditions refer to the temperature increase

$$\vartheta = \bar{\vartheta} \quad \text{on} \quad \partial B_t^{D_\Theta}, \quad \mathbf{q} \cdot \mathbf{n} = \bar{q}_N \qquad \text{on} \quad \partial B_t^{N_\Theta} \tag{3.34}$$

Analogous to the mechanical differential equation, the boundary is uniquely divided into a Dirichlet and a Neumann part. Since the differential equation is of 1st order in time, only the temperature change at the beginning has to be specified

$$\vartheta\,(t = t_0) = \vartheta_0 \quad \text{in} \quad B_t. \tag{3.35}$$

3.3.3 *Interface Effects*

If the printing process is based on melting, the material changes its aggregate state when a critical temperature is reached. In the case of an isothermal phase change, the interface corresponds to a surface in the body. This effect can be observed with pure materials. With crystalline alloys and materials possessing a continuous mixing, there is a temperature range at which the phase change takes place. This range is limited at the transition from solid to liquid by the solidus and liquidus temperature and the transition from liquid to gaseous by the bubble and dew temperature. To change the aggregate state, additional thermal energy is required which is accompanied by a change in volume. The additional energy for the phase change is described by the latent heat. A detailed theoretical treatment on the phase change of metals can be found in Glicksman (2011).

Mathematically, the process for isothermal phase change can be described by the Stefan condition. The change of the heat flow in the direction of the normal vector at the boundary layer $\mathbf{n}_{Phase}$ corresponds to the latent heat multiplied by the velocity of the boundary layer in the direction of phase change

$$\left(\mathbf{q}_s - \mathbf{q}_f\right) \cdot \mathbf{n}_{Phase} = L_{\mathbf{F}} \mathbf{v} \cdot \mathbf{n}_{Phase} \quad \text{if} \quad \Theta = \Theta_{Phase}. \tag{3.36}$$

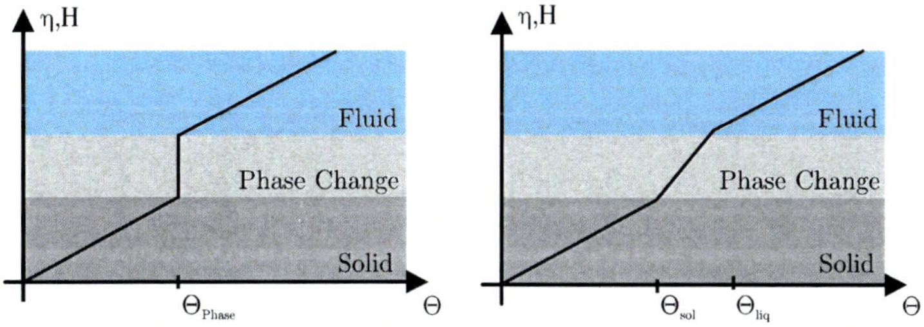

Fig. 3.2 Schematic change of enthropy and enthalpy due to temperature. (Left) Isothermal (right) non-isothermal phase change

This equation can also be derived from the 1st law of thermodynamics (3.26) if the height of a continuous layer approaches zero. An analytical solution for the 1-dimensional case can be found in Carslaw and Jaeger (1959) and numerical formulations in Wang et al. (1993) or Ji et al. (2002).

If the phase change can be assumed to be isobaric and reversible, the specific entropy is also proportional to the specific enthalpy H (Glicksman 2011)

$$\dot{\eta} = \frac{\dot{H}}{\Theta}. \tag{3.37}$$

If this relation holds, the 1st law of thermodynamics can be formulated alternatively in terms of the specific enthalpy

$$\rho \dot{H} = D_{int} - \operatorname{div} \mathbf{q} + \rho r. \tag{3.38}$$

The isothermal phase change causes a jump in enthalpy at the critical temperature. In case of non-isothermal phase change the enthalpy increases in the range between the transition temperatures (Fig. 3.2). The approach of Eyres et al. (1946) describes the latent heat in the critical temperature range solely as a function of the temperature

$$H = c_{\mathbf{F}}\Theta + L(\Theta). \tag{3.39}$$

A numerical implementation of this approach can be found, for example, in Voller and Prakash (1987), Chiumenti et al. (2010) or Rolph and Bathe (1982). Since the latent heat depends only on the temperature in this method, it can also be integrated into the specific heat capacity. This approach is only defined for non-isothermal phase changes and called apparent heat model (Bonacina et al. 1973)

$$c_{app} = \begin{cases} c_{s,f} & \Theta < \Theta_{sol,bub} \\ c_{inter} = \frac{\int_{\Theta_{s,f}}^{\Theta_{f,g}} c(\Theta)\mathrm{d}\Theta + L_{\mathbf{F}}^{mel,vap}}{\Theta_{liq,dew} - \Theta_{sol,bub}} & \Theta_{sol,bub} \leq \Theta \leq \Theta_{liq,dew} \\ c_{f,g} & \Theta > \Theta_{liq,dew} \end{cases}. \tag{3.40}$$

Alternatively, this phase change model can be integrated directly into the specific strain energy function (de Saracibar et al. 2001).

Remark In the presented modeling approaches, the volume change that occurs during a phase change is neglected. At the transition from solid to liquid, this influence is small. However, if the state changes from liquid to gaseous, this effect is evident. For example, 1 l water expands to a 1700 l gaseous state after vaporization.

3.3.4 Weak Form

The weak form is obtained by multiplying the 1st law of thermodynamics with a scalar test function τ. Using Gauss's integral theorem, the relationship can again be divided into a volume and a surface term. For the approach from (3.38) the weak form results to

$$G(\tau, \Theta) = \int_{B_t} \left(\tau[\rho \dot{H} - D_{int} - \rho r] - \operatorname{grad}\tau \cdot \mathbf{q}\right) \, \mathrm{d}v + \int_{\partial B_t^{N\Theta}} \tau q_N \, \mathrm{d}a = 0. \tag{3.41}$$

The order of the original differential equation is reduced by one and the Neumann boundary conditions are directly included. Equation (3.41) must only fulfill the Dirichlet boundary conditions. At these points, the test function must be zero because the derivatives are not known, i.e. $\tau = 0$ on $\partial B_t^{D\Theta}$.

3.4 Chemical Differential Equation

Instead of a phase change from liquid to solid, the material hardens during photopolyermization as a result of a chemical reaction. In symbolic notation, the starting material (A) reacts with the crosslinker (B) to form the product (C)

$$A + B \rightarrow C. \tag{3.42}$$

A and B are also called reactants. Using stoichiometry, the reaction can be described by relative mass fractions

$$\frac{m_A(t)}{m} + \frac{m_B(t)}{m} + \frac{m_C(t)}{m} = 1 \tag{3.43}$$

where $m_i(t)$, $i = A, B, C$ indicate the mass fractions of the individual substances at time t. During the reaction the total mass is constant

$$m = m_A(t) + m_B(t) + m_C(t) = m_A(0) + m_B(0) = const. \tag{3.44}$$

The quantities $m_i(0)$ correspond to the mass of each constituent at the beginning of the reaction at time $t = t_0$. Alternatively, the conservation of mass can be described using a scalar variable $\alpha(t)$ which represents the mass fraction of the product

$$\alpha(t) = \frac{m_C(t)}{m}. \tag{3.45}$$

With this quantity and (3.44) an approach for the change of the mass fractions can be formulated

$$\frac{m_A(t)}{m} = \frac{m_A(0)}{m}\,[1 - \alpha(t)]\,, \quad \frac{m_B(t)}{m} = \left(1 - \frac{m_A(0)}{m}\right)[1 - \alpha(t)]\,. \tag{3.46}$$

The process variable can only be in the range between 0 and 1, with $\alpha(t = t_0) = 0$ and $\alpha(t = t_\infty) = 1$. The constitutive equation for α must be determined by experiments. Further literature on reaction kinetics can be found in Schmidt (1998) and a detailed derivation of the above equations in Lion and Höfer (2007). A description of the process kinetics of rubber vulcanization is given in André and Wriggers (2005).

If the chemical reaction describes a curing process, $\alpha(t)$ can also be defined as the degree of cure. This value is often determined by the evolution of the reaction heat $H(t)$ divided by its maximum value H_∞

$$\alpha(t) = H(t)/H_\infty. \tag{3.47}$$

An overview of different approaches to model curing can be found in Lion and Höfer (2007). The model in Sourour and Kamal (1976) can be applied to describe the chemical reactions during photopolymerization. In this case, the cross-linking is approximated by an Arrhenius approach

$$\dot{\alpha} = \left[A_1 \exp\left(-\frac{B_1}{\Theta}\right) + A_2 \exp\left(-\frac{B_2}{\Theta}\right)\alpha^m\right](1 - \alpha)^n. \tag{3.48}$$

Curing describes an exothermic reaction. Its modeling can be based on the assumption that the specific strain energy function also depends on the degree of cure $\psi := f(\mathbf{F}, \Theta, \alpha)$ (Lion and Höfer 2007 or Landgraf et al. 2014). The 2nd law of thermodynamics indicates a dissipation due to the chemical reaction which corresponds to the exothermic heat. For elastic material behavior, the strain energy function and (3.48) must be chosen in such a way that the following condition is always fulfilled

$$D_{int} = -\rho \frac{\partial \psi}{\partial \alpha} \dot{\alpha} \geq 0. \tag{3.49}$$

Since the specific entropy depends on the degree of cure, an additional term results from the derivative of η with respect to time. Hence, the 1st law of thermodynamics is supplemented by two contributions due to chemical reactions where one of them is included in the dissipation

$$\rho L_{\mathbf{F}}\,\mathrm{tr}\,\mathbf{d} + \rho c_{\mathbf{F}}\dot{\Theta} - \rho \frac{\partial^2 \psi}{\partial \Theta \partial \alpha}\dot{\alpha} = D_{int} - \mathrm{div}\,\mathbf{q} + \rho r. \tag{3.50}$$

The third term on the left-hand side corresponds to the supply of heat at constant temperature and deformation which is necessary to increase the degree of cure by 1.

When modeling the curing process, certain material parameters usually depend on α. When an elastic structure is loaded, strains and stresses occur in the material. If no further load is applied, but the cross-linking is set in motion, additional stresses would result from the change of the material parameters. This, however, is unphysical. Therefore the assumption is made that the change of the curing process can change material parameters but not have any influence on the stress (e.g. André and Wriggers 2005). This leads to the constraint

$$\frac{\partial \boldsymbol{\sigma}}{\partial \alpha} \overset{!}{=} 0. \tag{3.51}$$

3.4.1 Initial Conditions

The chemical differential equation only describes an initial value problem. Hence only the degree of cure at the beginning ($t = t_0$) is to be specified

$$\alpha\,(t = t_0) = \alpha_0 \quad \text{in} \quad \mathcal{B}_t. \tag{3.52}$$

References

M. André, P. Wriggers, Thermo-mechanical behaviour of rubber materials during vulcanization. Int. J. Solids Struct. **42**(16–17), 4758–4778 (2005)

H.D. Baehr, S. Kabelac, *Thermodynamik*, 16th edn. (Springer, 2016)

C. Bonacina, G. Comini, A. Fasano, M. Primicerio, Numerical solution of phase-change problems. Int. J. Heat Mass Trans. **16**(10), 1825–1832 (1973)

A.N. Brooks, T.J.R. Hughes, Streamline upwind/Petrov-Galerkin formulations for convection dominated flows with particular emphasis on the incompressible Navier-Stokes equations. Comput. Methods Appl. Mech. Eng. **32**(1–3), 199–259 (1982)

H.S. Carslaw, J.C. Jaeger, *Conduction of Heat in Solids* (Oxford Science Publications, 1959)

M. Chiumenti, M. Cervera, A. Salmi, C.A. De Saracibar, N. Dialami, K. Matsui, Finite element modeling of multi-pass welding and shaped metal deposition processes. Comput. Methods Appl. Mech. Eng. **199**(37–40), 2343–2359 (2010)

B.D. Coleman, M.E. Gurtin, Thermodynamics with internal state variables. J. Chem. Phys. **47**(2), 597–613 (1967)

C.A. de Saracibar, M. Cervera, M. Chiumenti, On the constitutive modeling of coupled thermomechanical phase-change problems. Int. J. Plast. **17**(12), 1565–1622 (2001)

A.C. Eringen, C.B. Kafadar, Polar field theories, in *Continuum Physics*, ed. by A.C. Eringen. Polar and Nonlocal Field Theories, vol. IV (Academic Press, New York, 1976), pp. 1–74

N.R. Eyres, D.R. Hartree, J. Ingham, R.J. Sarjant, J.B. Wagstaff, The calculation of variable heat flow in solids. Phil. Trans. R. Soc. Lond. A **240**(813), 1–57 (1946)

M.E. Glicksman, *Principles of Solidification: An Introduction to Modern Casting and Crystal Growth Concepts* (Springer, New York, 2011)

P. Haupt, *Continuum Mechanics and Theory of Materials*, 2nd edn. (Springer, Berlin, Heidelberg, Wien, 2002)

G.A. Holzapfel, *Nonlinear Solid Mechanics A Continuum Approach for Engineering* (Wiley, Chichester, 2000)

T.R.J. Hughes, *The Finite Element Method* (Prentice Hall, Englewood Cliffs, NJ, 1987)

H. Ji, D. Chopp, J.E. Dolbow, A hybrid extended finite element/level set method for modeling phase transformations. Int. J. Num. Methods Eng. **54**(8), 1209–1233 (2002)

A. Krawietz, *Materialtheorie* (Springer, Berlin, Heidelberg, 1986)

L.D. Landau, E.M. Lifshitz, *Fluid mechanics: Volume 6 of Course of Theoretical Physics*, 2rd edn. (Pergamon Press, 1987)

R. Landgraf, M. Rudolph, R. Scherzer, J. Ihlemann, Modelling and simulation of adhesive curing processes in bonded piezo metal composites. Comput. Mech. **54**(2), 547–565 (2014)

A. Lion, P. Höfer, On the phenomenological representation of curing phenomena in continuum mechanics. Arch. Mech. **59**(1), 59–89 (2007)

D. Lüdecke, C. Lüdecke. *Thermodynamik: Physikalisch-chemische Grundlagen der thermischen Verfahrenstechnik* (Springer, 2013)

L.E. Malvern, *Introducation to the Mechanics of a Continuous Medium* (Prentice-Hall Inc, Englewood Cliffs, NJ, 1969)

J. Marsden, T.J.R. Hughes, *Mathematical Foundations of Elasticity* (Prentice-Hall Inc, Englewood Cliffs, 1983)

C. Miehe, *Zur numerischen Behandlung thermomechanischer Prozesse*. Ph.D. thesis, Universität Hannover, Germany (1988)

W.D. Rolph, K.J. Bathe, An efficient algorithm for analysis of nonlinear heat transfer with phase changes. Int. J. Num. Methods Eng. **18**(1), 119–134 (1982)

L.D. Schmidt, *The Engineering of Chemical Reactions*, vol. 2 (Oxford University Press, USA, 1998)

H.R. Schwarz, *Methode Der Finiten Elemente: Eine Einführung Unter Berücksichtigung Der Rechenpraxis* (Teubner, Stuttgart, 1991)

M. Silhavy, *The Mechanics and Thermodynamics of Continuous Media* (Springer, Berlin, Heidelberg, Wien, 1997)

S. Sourour, M.R. Kamal, Differential scanning calorimetry of epoxy cure: isothermal cure kinetics. Thermochimica Acta **14**(1–2), 41–59 (1976)

C. Truesdell, W. Noll, *The Non-Linear Field Theories of Mechanics*, 3rd edn. (Springer, Berlin, Heidelberg, New York, 2004)

C. Truesdell, R. Toupin. The classical field theories, in *Principles of Classical Mechanics and Field Theory/Prinzipien der Klassischen Mechanik und Feldtheorie* (Springer, 1960), pp. 226–858

V.R. Voller, C. Prakash, A fixed grid numerical modelling methodology for convection-diffusion mushy region phase-change problems. Int. J. Heat Mass Trans. **30**(8), 1709–1719 (1987)

S.L. Wang, R.F. Sekerka, A.A. Wheeler, B.T. Murray, S.R. Coriell, R.J. Braun, G.B. McFadden, Thermodynamically-consistent phase-field models for solidification. Physica D: Nonlinear Phenomena **69**(1–2), 189–200 (1993)

P. Wriggers, *Nonlinear Finite Element Methods* (Springer Science & Business Media, 2008)

Chapter 4
Meshfree Discretization Schemes

The simulation of Additive Manufacturing is based on coupled differential equations. Besides precise models for all phenomena, accurate solution schemes for differential equations are a prerequisite for realistic reproduction. Since analytical solutions exist only for certain special cases (Braess 2007), nowadays mainly numerical solution schemes are employed. The first step in these methods is a spatial and temporal discretization of the differential equations. In most cases, an algebraic equation must be solved to determine an approximate solution. Often the wrong results of a simulation are based on errors in the spatial discretization. Therefore, the emphasis is on this point. Time integration or solution schemes are outlined only in a subordinate way. In the case of meshfree methods, both topics are treated analogously to the Finite Element Method. More information on temporal discretization and on solution schemes can be found either in the literature on the FEM e.g. in Hughes (1987), Bathe (2006), Belytschko et al. (2000b), Wriggers (2008), Simo and Tarnow (1992) or in Wood (1990).

For spatial discretization various approaches are available. These can be divided into mesh-based and meshfree methods, whereas mixed variants also exist. In addition, the domain can be divided into particles or nodes plus integration points. In the former, the system is characterized as a one-point and in the latter as a two-point method. Moreover, there exist different concepts that represent the starting point for spatial discretization (Fig. 4.1).

Many simulations in an industrial environment are still based on the Finite Element Method. This scheme is particularly convincing due to an accurate solution of differential equations in many cases. For the calculation of solid deformations, the Lagrangian description of the differential equation is used to be able to directly analyze the actual deformations. Flow simulations lead to very large deformations. The elements can distort to such an extent that an accurate solution is no longer possible. For very large deformations, the Eulerian description is mostly preferred. This leads to an advection-diffusion term in the differential equation, which can cause numerical difficulties (Gresho and Sani 1998; Brooks and Hughes 1982). Further problems arise when modeling free surfaces or boundary layers. This requires addi-

C. Weißenfels, *Simulation of Additive Manufacturing using Meshfree Methods*,
Lecture Notes in Applied and Computational Mechanics 97,
https://doi.org/10.1007/978-3-030-87337-0_4

Concept	Neighborhood	Subdivision
Weak form	Meshfree	Two points
Weak form based on differences	Mesh-based	One point
Strong form	Mixed	
Reduced order strong form		

Fig. 4.1 Classification of spatial discretization schemes based on concept, neighborhood and subdivision

tional algorithms, such as Volume of Fluid (Hirt and Nichols 1981) or Level Sets (Osher and Sethian 1988; Osher and Fedkiw 2001).

Meshfree methods, on the other hand, also allow flow simulations in a Lagrangian description. Free surfaces, boundary layers, and the fusion of particles can be directly represented. On the other hand, meshfree methods usually cannot guarantee sufficient accuracy without additional treatment. This deficiency still hinders the increased use of these schemes in an industrial environment.

The main focus of this chapter represents the investigation of all requirements for spatial discretization schemes to enable an accurate solution of differential equations. Especially the divergence operator is studied in more detail since this contribution is mostly responsible for inaccuracies.

All investigations are carried out exclusively for the mechanical differential equation represented in the Lagrangian description and with respect to the current configuration. However, the individual statements can be directly transferred to other forms.

First, the most important terms are introduced and different discretization concepts are presented. The solution of nonlinear differential equations, as they occur in the modeling of Additive Manufacturing processes, requires special treatment. Often, the Newton-Raphson algorithm is used in this case, which is introduced in Sect. 4.3. Before listing all requirements for spatial discretization schemes in Sect. 4.5, possible reference configurations are defined in Sect. 4.4. Finally, common shape functions in the field of meshfree methods are presented in Sect. 4.6.

4.1 Points and Neighborhoods

In all solution methods, the domain is represented by a finite number of points. A point P defines a position in space that is localized by a coordinate but has no mass. The FEM distinguishes between integration points and nodes. At the integration points, the integral of the weak form is evaluated. The nodes correspond to the connection points between the individual elements. At these points, the values of the approximate solution and its time derivatives are determined. Quantities such as

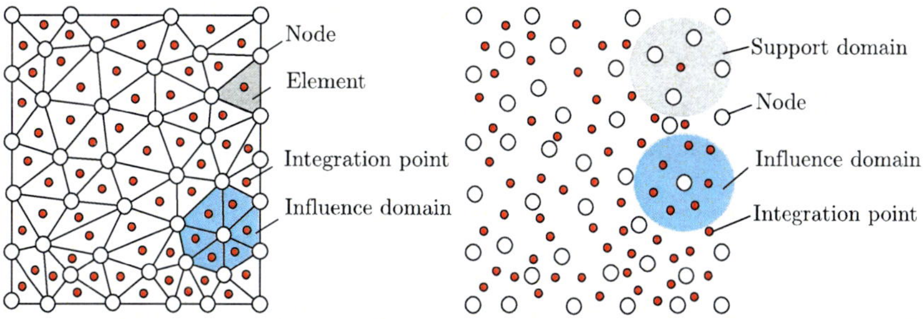

Fig. 4.2 Two-point methods: Subdivision of the domain into nodes and integration points. Identification of support and influence domain. (Left) Mesh-based methods. (Right) Meshfree methods

the stress, the strain, or the density are calculated at the integration points. In many meshfree methods, which also represent the domain by two different types of points, the same terminology is mostly used. Particles correspond to points that possess a mass and at which all quantities are evaluated. Thus, the domain is subdivided into only one type of point.

In all discretization methods, a neighborhood is assigned to each point. A neighborhood of P defines the sum of all points in the direct environment including P. In a two-point method, the nodes in the environment of an integration point or the reverse relation do not correspond to a neighborhood. For this reason, other terms have been established. In the FEM, the environment of an integration point corresponds to the element, where several integration points share the same environment. On the other hand, an environment consisting of integration points can be assigned to each node (Fig. 4.2), although this consideration usually plays a rather subordinate role in the FEM. Analogous definitions can be introduced for meshfree methods that divide the domain into two distinct types of points. In contrast to the FEM, a search algorithm determines the environment. The sum of nodes belonging to an integration point defines the support domain. The influence domain includes the integration points in the environment of a node. For meshfree methods that represent the domain by individual particles (Fig. 4.3), the support and influence domains coincide. In addition, the criteria for a neighborhood are met. For this reason, only this term is here used in the case of one-point methods.

Fig. 4.3 One-point meshfree method: Subdivision of the domain into particles and determination of the neighborhood by means of a search algorithm

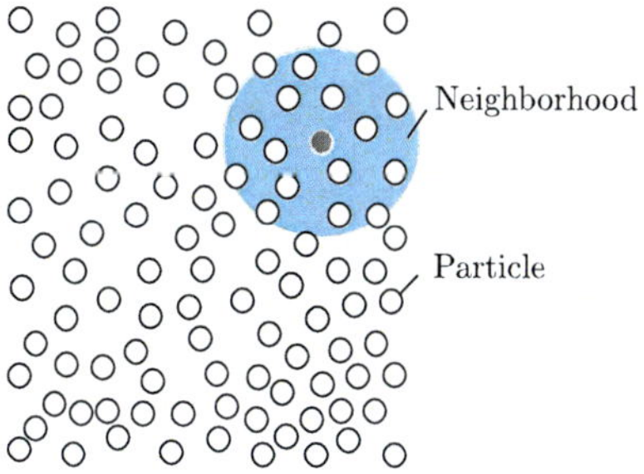

4.2 Discretization Concepts

Common to all methods, which are presented here, is the definition of the approximate solution in the neighborhood of a point by shape or basis functions N_I and coefficients $\mathbf{a}_I$

$$\mathbf{u}(\mathbf{x}) = \sum_{I=1}^{N_\mathbf{x}} N_I(\mathbf{x})\mathbf{a}_I. \tag{4.1}$$

The abbreviation $N_\mathbf{x}$ corresponds to the number of points in the neighborhood $H_\mathbf{x}$ of $\mathbf{x}$. The shape functions are defined in advance. Instead of the approximate solution, only the unknown coefficients have to be determined. In the FEM, the approximate solution is also called the trial function. To better show the analogies and differences, the term trial function is mainly used in the following.

All discretization methods lead in the final form of the mechanical differential equation to the condition that the force $\mathbf{f}_I$ at inner nodes or particles must be equal to zero and for nodes or particles at the Neumann boundary equal to the corresponding force $\mathbf{f}_I^n$

$$\mathbf{f}_I = \begin{cases} 0 & \text{for inner nodes/particles} \\ \mathbf{f}_I^n & \text{for boundary nodes/particles} \end{cases}. \tag{4.2}$$

The force $\mathbf{f}_I$ breaks down into a fraction from the dead load $\mathbf{f}_I^g$, the inertia term $\mathbf{f}_I^t$, and the internal force due to the state of stress inside the material $\mathbf{f}_I^s$

$$\mathbf{f}_I = \mathbf{f}_I^t + \mathbf{f}_I^g + \mathbf{f}_I^s. \tag{4.3}$$

The subscript corresponds to the location where the quantity is evaluated, i.e. $\mathbf{f}_I = \mathbf{f}\left(\mathbf{x}_I\right)$. Sought are the displacements or more precisely the coefficients that satisfy the equilibrium of forces. All discretization schemes can be classified into four different categories. At this stage, it does not matter whether the domain is represented by single particles or nodes plus integration points.

Remark. Often the force from the dead load and the force due to the Neumann boundary condition are combined as the external force $\mathbf{f}_I^{ext}$. Thus, it can also be required that at each particle the internal force $\mathbf{f}_I^{int} = \mathbf{f}_I^s$ plus $\mathbf{f}_I^t$ must correspond to the external force.

4.2.1 Weak Form

The Optimal Transportation Meshfree, the Element Free Galerkin, the Moving Least-Square Reproducing Kernel, and the Reproducing Kernel Particle method, respectively, are based, like the FEM, on the weak form of the differential equation (3.21)

$$G(\boldsymbol{\eta}, \mathbf{u}) = \int_{B_t} \left[\operatorname{grad} \boldsymbol{\eta} \cdot \boldsymbol{\sigma} + \rho \boldsymbol{\eta} \cdot (\ddot{\mathbf{u}} - \bar{\mathbf{b}}) \right] \, dv - \int_{\partial B_t^{N_\mathbf{u}}} \boldsymbol{\eta} \cdot \bar{\mathbf{t}} \, da = 0.$$

Discretization schemes, which are based on the weak form, divide the domain either into single particles or nodes plus integration points. Therefore, the discrete equations of both variants are briefly presented.

Two-point method. In this case, the integral in (3.21) is replaced by a sum over all integration points

$$\sum_{p=1}^{n_{gp}} \left(\operatorname{grad} \boldsymbol{\eta}_p \cdot \boldsymbol{\sigma}_p + \rho_p \boldsymbol{\eta}_p \cdot (\ddot{\mathbf{u}}_p - \bar{\mathbf{b}}) \right) v_p = \sum_{p=1}^{n_{gp}^s} \boldsymbol{\eta}_p \cdot \bar{\mathbf{t}}_p a_p. \tag{4.4}$$

The abbreviation n_{gp} defines the total number of integration points in the domain, whereas n_{gp}^s corresponds to the number of evaluation points at the surface. From the discretization of the test function

$$\boldsymbol{\eta}_p = \sum_{I=1}^{N_p^{sup}} N_I(\mathbf{x}_p) \, \mathbf{b}_I, \qquad \operatorname{grad} \boldsymbol{\eta}_p = \sum_{I=1}^{N_p^{sup}} \mathbf{b}_I \otimes \frac{\partial N_I(\mathbf{x}_p)}{\partial \mathbf{x}_p} \tag{4.5}$$

the weak form can be formulated in dependence on the components $\mathbf{b}_I$

$$\begin{aligned} \sum_{p}^{n_{gp}} \sum_{I=1}^{N_p^{sup}} \left(\mathbf{b}_I \otimes \frac{\partial N_I(\mathbf{x}_p)}{\partial \mathbf{x}_p} \cdot \boldsymbol{\sigma}_p + \rho_p N_I(\mathbf{x}_p) \, \mathbf{b}_I \cdot (\mathbf{u}_p - \bar{\mathbf{b}}) \right) v_p \\ = \sum_{p=1}^{n_{gp}^s} \sum_{I=1}^{N_p^{sup,s}} N_I(\mathbf{x}_p) \, \mathbf{b}_I \cdot \bar{\mathbf{t}}_p a_p. \end{aligned} \tag{4.6}$$

Besides the number of nodes in the support domain N_p^{sup}, the number of nodes at the surface belonging to an integration point at the boundary $N_p^{sup,s}$ is needed. The sum signs can be interchanged. Thus, the discrete form can also be represented as a sum over all nodes n

$$\sum_{I=1}^{n} \mathbf{b}_I \cdot \sum_{p=1}^{N_I^{inf}} \left(\boldsymbol{\sigma}_p \frac{\partial N_I(\mathbf{x}_p)}{\partial \mathbf{x}_p} + N_I(\mathbf{x}_p) \, \rho_p \, (\ddot{\mathbf{u}}_p - \bar{\mathbf{b}}) \right) v_p = \sum_{I=1}^{n^s} \mathbf{b}_I \cdot \sum_{p=1}^{N_I^{inf,s}} N_I(\mathbf{x}_p) \, \bar{\mathbf{t}}_p a_p. \tag{4.7}$$

At the Neumann boundary, for simplicity, we can also sum only over nodes n^s that lie at the corresponding surface. The abbreviation N_I^{inf} denotes the integration points in the influence domain of the node I and $N_I^{inf,s}$ include the corresponding number of points at the surface. Since the coefficients of the test function can be chosen arbitrarily and thus not equal to zero, the following must hold for each node

$$\sum_{p=1}^{N_I^{inf}} \left(\sigma_p \frac{\partial N_I\left(\mathbf{x}_p\right)}{\partial \mathbf{x}_p} + N_I\left(\mathbf{x}_p\right) \rho_p \left(\ddot{\mathbf{u}}_p - \bar{\mathbf{b}}\right) \right) v_p = \sum_{p=1}^{N_I^{inf,s}} N_I\left(\mathbf{x}_p\right) \bar{\mathbf{t}}_p a_p. \tag{4.8}$$

The first term on the left-hand side determines the contribution of the internal force due to the state of stress in the material

$$\mathbf{f}_I^s = \sum_{p=1}^{N_I^{inf}} \sigma_p \frac{\partial N_I\left(\mathbf{x}_p\right)}{\partial \mathbf{x}_p} v_p. \tag{4.9}$$

The second part leads to the inertia force and the third term to the force due to the dead load

$$\mathbf{f}_I^t = \sum_{p=1}^{N_I^{inf}} N_I\left(\mathbf{x}_p\right) \rho_p \ddot{\mathbf{u}}_p v_p, \quad \mathbf{f}_I^g = -\sum_{p=1}^{N_I^{inf}} N_I\left(\mathbf{x}_p\right) \rho_p \bar{\mathbf{b}} v_p. \tag{4.10}$$

The expression on the right hand side results in the contribution of the force due to the Neumann boundary condition

$$\mathbf{f}_I^n = \sum_{p=1}^{N_I^{inf,s}} N_I\left(\mathbf{x}_p\right) \bar{\mathbf{t}}_p a_p. \tag{4.11}$$

One-point method. Analogous relations can also be given for the case that the domain is represented by single particles. The integral is replaced by the sum over all particles n or over all particles n_s at the surface with $\bar{\mathbf{t}} \neq 0$

$$\sum_{J=1}^{n} \left(\operatorname{grad} \boldsymbol{\eta}_J \cdot \boldsymbol{\sigma}_J + \rho_J \boldsymbol{\eta}_J \cdot \left(\ddot{\mathbf{u}}_J - \bar{\mathbf{b}}\right)\right) v_J = \sum_{J=1}^{n_s} \boldsymbol{\eta}_J \cdot \bar{\mathbf{t}}_J a_J. \tag{4.12}$$

In the discretization of the test function, only the individual values at the particles are considered

$$\boldsymbol{\eta}_J = \sum_{I=1}^{N_J} N_I\left(\mathbf{x}_J\right) \mathbf{b}_I, \qquad \operatorname{grad} \boldsymbol{\eta}_J = \sum_{I=1}^{N_J} \mathbf{b}_I \otimes \frac{\partial N_I\left(\mathbf{x}_J\right)}{\partial \mathbf{x}_J}. \tag{4.13}$$

The abbreviation N_J in the sum sign corresponds to the number of particles in the neighborhood of the particle J. The individual force components are calculated in an analogous way to the two-point method. The internal force due to the state of stress in the material results from the evaluation of the individual quantities in its neighborhood

$$\mathbf{f}_I^s = \sum_{J=1}^{N_I} \sigma_J \frac{\partial N_I\left(\mathbf{x}_J\right)}{\partial \mathbf{x}_J} v_J. \tag{4.14}$$

The second term in (4.12) leads to the inertial force and the third term to the force due to the dead load

$$\mathbf{f}_I^t = \sum_{J=1}^{N_I} N_I(\mathbf{x}_J)\,\rho_J \ddot{\mathbf{u}}_J v_J, \quad \mathbf{f}_I^g = -\sum_{J=1}^{N_I} N_I(\mathbf{x}_J)\,\rho_J \bar{\mathbf{b}} v_J. \tag{4.15}$$

Assuming a Kronecker-δ property of the shape function, both forces can directly be formulated in dependence on the mass at the corresponding particle

$$\mathbf{f}_I^t = m_I \ddot{\mathbf{u}}_I, \quad \mathbf{f}_I^g = -m_I \bar{\mathbf{b}}, \quad \text{with} \quad m_I = \rho_I v_I. \tag{4.16}$$

When calculating the force due to the Neumann boundary condition, it is sufficient to sum only over the particles N_I^s at the corresponding boundary

$$\mathbf{f}_I^n = \sum_{J=1}^{N_I^s} N_I(\mathbf{x}_J)\,\bar{\mathbf{t}}_J a_J. \tag{4.17}$$

The approximate solution must also fulfill the Dirichlet boundary conditions, i.e. $\mathbf{u}_I = \bar{\mathbf{u}}(\mathbf{x}_I)$ on $\partial B_t^{D_\mathrm{u}}$.

Remark. If the same shape functions are used for the trial and the test function, a Bubnov-Galerkin method is present. In the other case, the scheme belongs to a Petrov-Galerkin method. As pointed out in textbooks on the FEM, e.g. Wriggers (2008), the stiffness matrix for conservative systems must be symmetric. However, this cannot be guaranteed for Petrov-Galerkin methods. Mostly, this approach is used to correct numerical instabilities resulting, for example, from the advection-diffusion term in the differential equations (Brooks and Hughes 1982).

4.2.2 Strong Form

As an alternative to the weak form, the strong form can serve as a starting point for the simulation. In the field of meshfree methods, this form is often used for the calculation of fluid flows. Although a formulation based on a two-point scheme is also possible, a one-point method is more suitable, since no integration over the domain occurs. For this reason, only the discretization of the differential equation for fluids, which is based on a subdivision of the domain into particles, is presented in the following. For the case of incompressible fluids, the divergence of the stress splits into a hydrostatic and a viscous part (3.13)

$$\operatorname{div} \boldsymbol{\sigma} = -\operatorname{grad} p + 2\eta \operatorname{div} \operatorname{grad}^{sym} \mathbf{v}.$$

By substituting into the strong form, the differential equation can be formulated directly in dependence on the two solution functions, the pressure p and the velocity $\mathbf{v}$ (3.14)

$$\rho \dot{\mathbf{v}} = -\operatorname{grad} p + 2\eta \operatorname{div} \operatorname{grad}^{sym} \mathbf{v} + \rho \bar{\mathbf{b}}.$$

When considering the strong form, the differential equation must be satisfied at each particle

$$\rho_I \dot{\mathbf{v}}_I + \operatorname{grad} p_I - 2\eta \operatorname{div} \operatorname{grad}^{sym} \mathbf{v}_I - \rho_I \bar{\mathbf{b}} = 0. \tag{4.18}$$

The discretization of the pressure gradient is similar to (4.13). But the viscous part needs the second derivative of the shape function

$$\begin{aligned} \operatorname{grad} p_I &= \sum_{J=1}^{N_I} p_J \frac{\partial N_J(\mathbf{x}_I)}{\partial \mathbf{x}_I}, \\ \operatorname{div} \operatorname{grad}^{sym} \mathbf{v}_I &= \frac{1}{2} \sum_{J=1}^{N_I} \left[\mathbf{v}_J \otimes \frac{\partial^2 N_J(\mathbf{x}_I)}{\partial^2 \mathbf{x}_I} + \frac{\partial^2 N_J(\mathbf{x}_I)}{\partial^2 \mathbf{x}_I} \otimes \mathbf{v}_J \right] : \mathbf{1}. \end{aligned} \tag{4.19}$$

The multiplication symbol : denotes the double contraction of two tensors and $\mathbf{1}$ corresponds to the unit tensor. The individual forces result from the multiplication with the volume. The contributions from the inertia term and from the dead load are equivalent to (4.16). The internal force at the particle is determined from (4.19)

$$\mathbf{f}_I^s = v_I \left[\sum_{J=1}^{N_I} p_J \frac{\partial N_J(\mathbf{x}_I)}{\partial \mathbf{x}_I} - \eta \left[\mathbf{v}_J \otimes \frac{\partial^2 N_J(\mathbf{x}_I)}{\partial^2 \mathbf{x}_I} + \frac{\partial^2 N_J(\mathbf{x}_I)}{\partial^2 \mathbf{x}_I} \otimes \mathbf{v}_J \right] : \mathbf{1} \right]. \tag{4.20}$$

The force $\mathbf{f}_I^n$ results from the satisfaction of the Neumann boundary condition $\bar{\mathbf{t}}$ at the corresponding particles

$$\boldsymbol{\sigma}_I \mathbf{n}_I = \bar{\mathbf{t}}(\mathbf{x}_I), \quad \text{on } \partial B_t^{N_\mathbf{u}}. \tag{4.21}$$

In addition, the requirement $\mathbf{u}_I = \bar{\mathbf{u}}(\mathbf{x}_I)$ must be satisfied at the Dirichlet boundary $\partial B_t^{D_\mathbf{u}}$.

4.2.3 Reduced Order Strong Form

Alternatively, the balance equation and the kinematic or constitutive quantities can be treated separately. The balance of momentum (3.3)

$$\rho \ddot{\mathbf{u}} = \operatorname{div} \boldsymbol{\sigma} + \rho \bar{\mathbf{b}}$$

requires only the first derivative. The same applies to the calculation of the kinematical relation. Thus, analogous to the weak form, the order of the highest derivative has decreased by one. Since no integration occurs, the discrete equations are only derived employing a one-point method. The balance equation must be fulfilled at each particle

$$\rho_I \ddot{\mathbf{u}}_I - \operatorname{div} \boldsymbol{\sigma}_I - \rho_I \bar{\mathbf{b}} = 0. \tag{4.22}$$

With the discrete form of the divergence operator, the individual force components in (4.3) can be specified more precisely

$$\operatorname{div} \boldsymbol{\sigma}_I = \sum_{J=1}^{N_I} \boldsymbol{\sigma}_J \frac{\partial N_J(\mathbf{x}_I)}{\partial \mathbf{x}_I}, \quad \mathbf{f}_I^s = -v_I \sum_{J=1}^{N_I} \boldsymbol{\sigma}_J \frac{\partial N_J(\mathbf{x}_I)}{\partial \mathbf{x}_I}. \tag{4.23}$$

The contributions from the dead load and the inertia term are exactly equal to (4.16). The force $\mathbf{f}_I^n$ is determined analogously to the previous section. Compared to the weak form the only difference in the calculation of the total force $\mathbf{f}_I$ is the determination of the internal force $\mathbf{f}_I^s$. Besides the multiplication with the volume and the exchanged order of the indices in the derivative of the shape function, the sign has also changed.

4.2.4 Weak Form Based on Differences

This type constitutes the basis for the correspondence formulation within Peridynamics. As will be shown in Chap. 6 it also leads to a typical formulation in SPH. Since both discretization schemes are based on a subdivision of the domain into individual particles, only this concept is considered. The gradient of the test function in (3.21) can alternatively be approximated from a Taylor series. The function η at the neighboring point $\mathbf{x}'$ of $\mathbf{x}$ can also be determined from the sum of derivatives multiplied by differences

$$\eta\left(\mathbf{x}'\right) = \eta\left(\mathbf{x}\right) + \operatorname{grad} \eta \left(\mathbf{x}' - \mathbf{x}\right) + \mathcal{O}\left(\left(\mathbf{x}' - \mathbf{x}\right)^2\right), \quad \operatorname{grad} \eta = \frac{\partial \eta}{\partial \mathbf{x}}. \tag{4.24}$$

The gradient of the test function can now be determined using the method of least squares. Only the linear part in the Taylor series is considered and the error in the integral over the whole neighborhood is minimized

$$J := \int_{H_\mathbf{x}} \omega\left(\mathbf{x}' - \mathbf{x}\right) \|\eta\left(\mathbf{x}'\right) - \eta\left(\mathbf{x}\right) - \operatorname{grad} \eta \left(\mathbf{x}' - \mathbf{x}\right)\|^2 \, \mathrm{d}v_{\mathbf{x}'} \quad \rightarrow \quad \min. \tag{4.25}$$

The Euclidean norm in the integral is additionally weighted by the function ω. The subscript at $\mathrm{d}v$ indicates the variable to be integrated. The gradient of the test function which minimizes the functional J results from the requirement that the derivative of J with respect to $\operatorname{grad}\boldsymbol{\eta}$ must be equal to zero

$$\frac{\partial J}{\partial \operatorname{grad}\boldsymbol{\eta}} \stackrel{!}{=} 0 \rightarrow \operatorname{grad}\boldsymbol{\eta} = \int_{H_{\mathbf{x}}} \omega\left(\mathbf{x}'-\mathbf{x}\right)\left[\boldsymbol{\eta}\left(\mathbf{x}'\right)-\boldsymbol{\eta}\left(\mathbf{x}\right)\right]\otimes\left(\mathbf{x}'-\mathbf{x}\right)\,\mathrm{d}v_{\mathbf{x}'}\mathbf{M}^{-1}\left(\mathbf{x}\right). \tag{4.26}$$

In the context of least squares, the tensor $\mathbf{M}$ is called moment tensor

$$\mathbf{M}\left(\mathbf{x}\right) = \int_{H_{\mathbf{x}}} \omega\left(\mathbf{x}'-\mathbf{x}\right)\left(\mathbf{x}'-\mathbf{x}\right)\otimes\left(\mathbf{x}'-\mathbf{x}\right)\,\mathrm{d}v_{\mathbf{x}'}. \tag{4.27}$$

More information on least squares can be found in Sect. 4.6. Substituting (4.26) into (3.21) and taking advantage of the symmetry of the moment tensor an alternative form for the virtual internal work can be obtained

$$\int_{B_t} \operatorname{grad}\boldsymbol{\eta}\cdot\boldsymbol{\sigma}\,\mathrm{d}v = \int_{B_t}\int_{H_{\mathbf{x}}} \left[\boldsymbol{\eta}\left(\mathbf{x}'\right)-\boldsymbol{\eta}\left(\mathbf{x}\right)\right]\cdot\mathbf{t}\left(\mathbf{x},\mathbf{x}'\right)\,\mathrm{d}v_{\mathbf{x}'}\,\mathrm{d}v_{\mathbf{x}}. \tag{4.28}$$

To simplify the notation the vector $\mathbf{t}$ is introduced

$$\mathbf{t}\left(\mathbf{x},\mathbf{x}'\right) = \omega\left(\mathbf{x}'-\mathbf{x}\right)\boldsymbol{\sigma}(\mathbf{x})\mathbf{M}^{-1}\left(\mathbf{x}\right)\left[\mathbf{x}'-\mathbf{x}\right]. \tag{4.29}$$

Since $\mathbf{t}\left(\mathbf{x},\mathbf{x}'\right)=0$, if $\mathbf{x}'\notin H_{\mathbf{x}}$ the integration can be extended over the entire domain. The right-hand side in (4.28) can be transformed by exchanging indices

$$\begin{aligned}
&\int_{B_t}\int_{B_t} \left[\boldsymbol{\eta}\left(\mathbf{x}'\right)-\boldsymbol{\eta}\left(\mathbf{x}\right)\right]\cdot\mathbf{t}\left(\mathbf{x},\mathbf{x}'\right)\,\mathrm{d}v_{\mathbf{x}'}\,\mathrm{d}v_{\mathbf{x}}\\
&=\int_{B_t}\int_{B_t} \boldsymbol{\eta}\left(\mathbf{x}\right)\cdot\mathbf{t}\left(\mathbf{x}',\mathbf{x}\right)\,\mathrm{d}v_{\mathbf{x}}\,\mathrm{d}v_{\mathbf{x}'} - \int_{B_t}\int_{B_t} \boldsymbol{\eta}\left(\mathbf{x}\right)\cdot\mathbf{t}\left(\mathbf{x},\mathbf{x}'\right)\,\mathrm{d}v_{\mathbf{x}'}\,\mathrm{d}v_{\mathbf{x}}\\
&=-\int_{B_t}\int_{B_t} \boldsymbol{\eta}\left(\mathbf{x}\right)\cdot\left[\mathbf{t}\left(\mathbf{x},\mathbf{x}'\right)-\mathbf{t}\left(\mathbf{x}',\mathbf{x}\right)\right]\,\mathrm{d}v_{\mathbf{x}'}\,\mathrm{d}v_{\mathbf{x}}.
\end{aligned} \tag{4.30}$$

Thus, the weak form can also be formulated in dependence on a difference of $\mathbf{t}$

$$\begin{aligned}
-\int_{B_t}\int_{H_{\mathbf{x}}} \boldsymbol{\eta}\left(\mathbf{x}\right)\cdot\left[\mathbf{t}\left(\mathbf{x},\mathbf{x}'\right)-\mathbf{t}\left(\mathbf{x}',\mathbf{x}\right)\right]\,\mathrm{d}v_{\mathbf{x}'}\,\mathrm{d}v_{\mathbf{x}} + \int_{B_t} \rho\boldsymbol{\eta}\cdot\left(\ddot{\mathbf{u}}-\bar{\mathbf{b}}\right)\,\mathrm{d}v\\
=\int_{\partial B_t^{N_{\mathbf{u}}}} \boldsymbol{\eta}\cdot\bar{\mathbf{t}}\,\mathrm{d}a.
\end{aligned} \tag{4.31}$$

In the discrete form, the sum over all particles n replaces the integral over the volume. The contributions from the dead load, the inertia term, and the force due to the Neumann boundary condition are exactly the same as in Sect. 4.2.1. Only the internal force due to the state of stress in the material is calculated differently. In the discrete form, the corresponding force can be further specified by substituting (4.29) into (4.31)

$$\mathbf{f}_I^s = -\sum_{J=1}^{N_I} \Big(\omega \left(\mathbf{x}_J - \mathbf{x}_I\right) \boldsymbol{\sigma}_I \mathbf{M}_I^{-1} \left[\mathbf{x}_J - \mathbf{x}_I\right] - \omega \left(\mathbf{x}_I - \mathbf{x}_J\right) \boldsymbol{\sigma}_J \mathbf{M}_J^{-1} \left[\mathbf{x}_I - \mathbf{x}_J\right] \Big) v_J v_I . \tag{4.32}$$

With the definition

$$\frac{\partial N_I \left(\mathbf{x}_J\right)}{\partial \mathbf{x}_J} = \omega \left(\mathbf{x}_I - \mathbf{x}_J\right) \mathbf{M}_J^{-1} \left(\mathbf{x}_I - \mathbf{x}_J\right) v_I \tag{4.33}$$

the inner force at a particle can be formulated in dependence on shape function derivatives

$$\mathbf{f}_I^s = -\sum_{J=1}^{N_I} \left[\boldsymbol{\sigma}_I \frac{\partial N_J \left(\mathbf{x}_I\right)}{\partial \mathbf{x}_I} v_I - \boldsymbol{\sigma}_J \frac{\partial N_I \left(\mathbf{x}_J\right)}{\partial \mathbf{x}_J} v_J \right] . \tag{4.34}$$

If the derivatives in the first term sum to zero

$$\sum_{J=1}^{N_I} \frac{\partial N_J \left(\mathbf{x}_I\right)}{\partial \mathbf{x}_I} = 0 \tag{4.35}$$

Equation (4.34) corresponds exactly to the contribution from the weak form (4.14). This is given if the shape functions satisfy the reproducing conditions in the derivatives of 0th order (4.65).

Remark. The concept leading to (4.26) is similar to the so-called peridynamic derivative formulated in Madenci et al. (2016) and Madenci et al. (2018).

4.3 Solution Schemes

After discretization, the coefficients **a** that satisfy the equilibrium of forces have to be computed. The contributions from each node or particle are normally summarized in a vector **R**. In the Finite Element Method, this process is called assembly. To show the analogy to the FEM, the same term and the corresponding operator is used. Thus, the unknown coefficients **a** can be solved from the condition $\mathbf{R} = 0$

$$\mathbf{R} \left(\mathbf{a}, \ddot{\mathbf{a}}\right) = \mathop{\mathsf{A}}_{I=1}^{n} \left(\mathbf{f}_I - \mathbf{f}_I^n\right) = 0. \tag{4.36}$$

For dynamic systems a time discretization is required. Afterward, the condition reduces to $\mathbf{R}\left(\mathbf{a}\right) = 0$. If a quasi-static case is present, or an implicit time integration scheme is used, then the unknowns are calculated from a system of linear equations. In the case of a linear differential equation, the vector **R** can be directly decomposed into a fraction resulting from the multiplication of the stiffness matrix $\mathbf{K}$ with $\mathbf{a}$ and a load vector $\mathbf{p}$

$$\mathbf{R}(\mathbf{a}) = 0, \quad \rightarrow \quad \mathbf{Ka} - \mathbf{p} = 0. \tag{4.37}$$

The vector $\mathbf{p}$ includes all contributions from the loading that are independent of the displacements. In this case, the coefficients $\mathbf{a}$ can be determined directly from (4.37). The equations resulting form the modeling of Additive Manufacturing processes are normally non-linear with respect to the solution functions. In these cases, the Newton-Raphson algorithm is often selected to compute the unknowns. This scheme determines the coefficients incrementally and is based on a Taylor series. If only the linear part is considered, the equilibrium condition at step $k+1$ can be calculated from already known quantities at step k and the unknown increment $\Delta\mathbf{a}$

$$\mathbf{R}\left(\mathbf{a}^{k+1}\right) = 0, \qquad \mathbf{R}\left(\mathbf{a}^{k+1}\right) = \mathbf{R}\left(\mathbf{a}^{k}\right) + \frac{\partial \mathbf{R}\left(\mathbf{a}^{k}\right)}{\partial \mathbf{a}} \Delta\mathbf{a}^{k+1}. \tag{4.38}$$

With the definition of the tangent matrix $\mathbf{K}$, the increment is computed solving a system of linear equations

$$\mathbf{K}(\mathbf{a}^{k})\Delta\mathbf{a}^{k+1} = -\mathbf{R}\left(\mathbf{a}^{k}\right), \qquad \mathbf{K}(\mathbf{a}^{k}) = \frac{\partial \mathbf{R}\left(\mathbf{a}^{k}\right)}{\partial \mathbf{a}}. \tag{4.39}$$

The vector $\mathbf{a}^{k+1}$ is determined directly from the increment

$$\mathbf{a}^{k+1} = \mathbf{a}^{k} + \Delta\mathbf{a}^{k+1}. \tag{4.40}$$

Different conditions are available as termination criteria for the Newton-Raphson algorithm, such as the Euclidean norm of the vector $\mathbf{R}$. In this case, the equilibrium is found if the norm falls below a chosen tolerance δ

$$\|\mathbf{R}\left(\mathbf{a}^{k+1}\right)\| = \sqrt{\mathbf{R}\left(\mathbf{a}^{k+1}\right) \cdot \mathbf{R}\left(\mathbf{a}^{k+1}\right)} \leq \delta. \tag{4.41}$$

This algorithm converges quadratically if the initial value is close to the solution we are looking for. The disadvantage of this method is the constant recalculation of the tangent matrix $\mathbf{K}$. Further information and alternative solution schemes for nonlinear equations can be found e.g. in Wriggers (2008) or Belytschko et al. (2000b).

Remark 1. In the Finite Element Method, assembly describes the assignment of local to global node numbers (Hughes 1987).

Remark 2. Explicit time integration schemes have the advantage that no tangent matrix has to be calculated and no inverse has to be determined. Especially these operations lead to high computational efforts when the number of degrees of freedom is very large. On the other hand, with these methods, the time step size usually has to be chosen much smaller to guarantee the stability of the integration scheme. In addition, no equilibrium is calculated in explicit methods. Depending on the task, the right solution method must therefore be carefully selected.

4.4 Reference Configurations

This chapter is limited to the formulation of the differential equations with respect to quantities of the current configuration. Furthermore, only the Lagrangian description is considered. The individual quantities are thus related to mass points. The mass point itself can be represented by a position vector with respect to different reference configurations. This section presents four possibilities (Fig. 4.4). For a more descriptive presentation, the dependence on the time or load step is also given in this section. The formulations are only presented for the discretization of the classical weak form. Additionally, it is assumed that the domain is subdivided into nodes and integration points. The procedure for other concepts is analogous.

4.4.1 Total Lagrangian Formulation

In the total Lagrangian formulation, all quantities are defined in dependence on the coordinate **X** of the initial configuration

$$\mathbf{f}_{I\,n+1} = \sum_{p=1}^{N_I^{inf}} \boldsymbol{\sigma}_{n+1}(\mathbf{X}_p)\,\frac{\partial N_I(\mathbf{X}_p)}{\partial \mathbf{x}_{p\,n+1}}\,v_{n+1}(\mathbf{X}_p). \tag{4.42}$$

For simple materials, the stress is a function of **F**, see also Sect. 3.2. The deformation gradient is calculated from the derivative of the shape function with respect

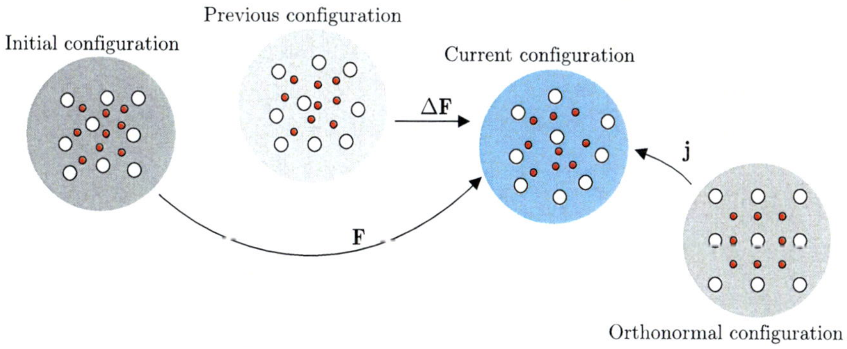

Fig. 4.4 Different Lagrangian reference configurations to formulate discretized differential equations

to the coordinates of the initial configuration and can thus be determined directly. The derivative of the shape function with respect to the current coordinates results from a multiplication with the inverse of the deformation gradient

$$\mathbf{F}_{p\,n+1} = \frac{\partial \mathbf{x}_{p\,n+1}}{\partial \mathbf{X}_p} = \mathbf{1} + \sum_{I=1}^{N} \mathbf{a}_{I\,n+1} \otimes \frac{\partial N_I(\mathbf{X}_p)}{\partial \mathbf{X}_p}, \quad \frac{\partial N_I(\mathbf{X}_p)}{\partial \mathbf{x}_{p\,n+1}} = \frac{\partial N_I(\mathbf{X}_p)}{\partial \mathbf{X}_p} \mathbf{F}^{-1}_{p\,n+1}. \tag{4.43}$$

The transfer from the initial to the current configuration is called push-forward. The inverse relation corresponds to a pull-back. A detailed treatise on this subject can be found in Marsden and Hughes (1983) or Miehe (1988). The current volume is calculated from the volume at the beginning v_0 multiplied by the determinant of the deformation gradient

$$v_{n+1}(\mathbf{X}_p) = \det \mathbf{F}_{p\,n+1}\, v_0(\mathbf{X}_p). \tag{4.44}$$

This formulation is not suitable for fluid flow simulations, since the closest neighboring points change during the calculation.

4.4.2 Current Lagrangian Formulation

To simulate very large deformations with meshfree methods, a redefinition of the neighborhood at each time step is advantageous. The force can also be calculated using quantities determined at points in the current neighborhood

$$\mathbf{f}_{I\,n+1} = \sum_{p=1}^{N^{inf}_{I\,n+1}} \boldsymbol{\sigma}_{n+1}(\mathbf{x}_{p\,n+1}) \frac{\partial N_I(\mathbf{x}_{p\,n+1})}{\partial \mathbf{x}_{p\,n+1}}\, v_{n+1}(\mathbf{x}_{p\,n+1}). \tag{4.45}$$

In this case, the deformation gradient is calculated from its inverse

$$\mathbf{F}^{-1}_{p\,n+1} = \mathbf{1} - \sum_{I=1}^{N^{sup}_{p\,n+1}} \mathbf{a}_{I\,n+1} \otimes \frac{\partial N_I(\mathbf{x}_{p\,n+1})}{\partial \mathbf{x}_{p\,n+1}}. \tag{4.46}$$

When using the Newton-Raphson algorithm, the dependence of the shape function on the current coordinate leads to an additional part in the linearization.

4.4.3 Updated Lagrangian Formulation

To avoid this extra effort in the linearization, an updated Lagrangian formulation can be used as an alternative. Here, all quantities refer to the coordinates at the previous time step t_n. This formulation also allows a simulation of very large deformations

$$\mathbf{f}_{I\,n+1} = \sum_{p=1}^{N_{I\,n}^{inf}} \boldsymbol{\sigma}_{n+1}(\mathbf{x}_{p\,n}) \, \frac{\partial N_I(\mathbf{x}_{p\,n})}{\partial \mathbf{x}_{p\,n+1}} \, v_{n+1}(\mathbf{x}_{p\,n}). \tag{4.47}$$

With the aid of the incremental deformation gradient

$$\Delta\mathbf{F}_{p\,n+1} = \frac{\partial \mathbf{x}_{p\,n+1}}{\partial \mathbf{x}_{p\,n}} = \mathbf{1} + \sum_{I=1}^{N_{p\,n}^{sup}} (\mathbf{a}_{I\,n+1} - \mathbf{a}_{I\,n}) \otimes \frac{\partial N_I(\mathbf{x}_{p\,n})}{\partial \mathbf{x}_{p\,n}} \tag{4.48}$$

the current deformation gradient and the current volume can be determined

$$\mathbf{F}_{p\,n+1} = \Delta\mathbf{F}_{p\,n+1}\mathbf{F}_{p\,n}, \qquad v_{p\,n+1} = \det \Delta\mathbf{F}_{p\,n+1} v_{p\,n}. \tag{4.49}$$

In (4.48), the identity $\mathbf{X}_p = \mathbf{x}_{p\,n} - \mathbf{a}_{p\,n}$ is used. The derivative of the shape function with respect to the current coordinates results from the push-forward

$$\frac{\partial N_I(\mathbf{x}_{p\,n})}{\partial \mathbf{x}_{p\,n+1}} = \frac{\partial N_I(\mathbf{x}_{p\,n})}{\partial \mathbf{x}_{p\,n}} \Delta\mathbf{F}_{p\,n+1}^{-1}. \tag{4.50}$$

4.4.4 Isoparametric Concept

Especially for mesh-based solution schemes, a formulation with respect to a configuration that has a regular structure is a good choice, see Weißenfels (2019). This also facilitates numerical integration and is called isoparametric concept in the FEM (Irons and Zienkiewicz 1968; Ergatoudis et al. 1968). The reference system is based on orthonormal coordinate vectors $\boldsymbol{\xi}$

$$\mathbf{f}_{I\,n+1} = \sum_{p=1}^{N_I^{inf}} \boldsymbol{\sigma}_{n+1}(\boldsymbol{\xi}_p) \, \frac{\partial N_I(\boldsymbol{\xi}_p)}{\partial \mathbf{x}_{p\,n+1}} \, v_{n+1}(\boldsymbol{\xi}_p). \tag{4.51}$$

The current derivatives result from the push-forward using the Jacobi tensor $\mathbf{j}$

$$\mathbf{j}_{p\,n+1} = \frac{\partial \mathbf{x}_{p\,n+1}}{\partial \boldsymbol{\xi}_p} = \mathbf{1} + \sum_{I=1}^{N^{sup}} \mathbf{a}_{I\,n+1} \otimes \frac{\partial N_I(\boldsymbol{\xi}_p)}{\partial \boldsymbol{\xi}_p}, \quad \frac{\partial N_I(\boldsymbol{\xi}_p)}{\partial \mathbf{x}_{n+1}} = \frac{\partial N_I(\boldsymbol{\xi}_p)}{\partial \boldsymbol{\xi}_p} \mathbf{j}_{p\,n+1}^{-1}. \tag{4.52}$$

For numerical integration, the Gauss method is usually employed. If the shape function corresponds of a polynomial, an exact integration can be ensured. The volume is calculated from the Jacobi determinant and a weighting W

$$v_{n+1}(\boldsymbol{\xi}_p) = \det \mathbf{j}_{p\,n+1} W_p. \tag{4.53}$$

The number of nodes in the influence and the support domain is defined at the beginning and is not redetermined during the simulation. Detailed information on the isoparametric concept can be found in the common literature on the Finite Element Method (Hughes 1987; Dhatt and Touzot 1984; Zienkiewicz and Taylor 1989).

4.5 Requirements on Spatial Discretization Schemes

For an accurate solution of the differential equation, a lot of requirements must be met. First, the solution has to exist and be unique. These aspects are part of functional analysis. A detailed treatment of this subject can be found, for example, in Vainberg (1964) and a proof in connection with nonlinear elasticity theory in Marsden and Hughes (1983). This work assumes the existence and the uniqueness of the solution.

Further issues arise in the numerical solution of differential equations. In this case, the solution function is approximated. Thus, it must be ensured that the given set of possible approximations contains the solution of the differential equation. This aspect is also part of functional analysis. If a solution exists and the approximation contains this solution, then it must be additionally ensured that the numerical procedure makes the finding of the solution possible. For this, several conditions must be fulfilled. Exactly this aspect will be discussed in more detail in this section.

Since these conditions must apply independently of the time or load step, an explicit specification of the time is omitted. In addition, the derivations are mainly based on the weak form. However, the individual conditions apply analogously to other methods. If differences are found, this is noted at the appropriate place. Before explaining the individual requirements on spatial discretization schemes, some important terms are defined first.

4.5.1 *Definitions*

In the literature on numerical solution schemes, the terms consistency, completeness, reproducing condition, convergence, stability, and locking appear frequently. To ease further understanding, these definitions are first explained in more detail.

Consistency, completeness, reproducing conditions. The term consistency is found in connection with the Finite Difference Method. This scheme solves the strong form of the differential equation on a fixed grid. The error that arises when the true solution $\mathbf{u}$ of the differential equation is substituted into the discrete formulation is investigated. The evaluation is based on the difference between the response of the discrete L_h and the continuous differential operator L. The difference must tend to zero as the resolution increases

$$\|L_h(\mathbf{u}) - L(\mathbf{u})\| \rightarrow 0 \quad \text{as} \quad h \rightarrow 0. \tag{4.54}$$

The grid spacing h is chosen as the measure of the resolution. The arbitrary shape and arrangement of the elements makes it difficult to prove consistency in the FEM. Instead, the completeness of the basis or shape functions is required. An approximation is complete within a neighborhood if the error of the approximation compared to a given function $\mathbf{u}$ approaches zero as the resolution increases (Belytschko et al. 2000b)

$$\|\mathbf{u}(\mathbf{x}) - \sum_{I=1}^{N_\mathbf{x}} N_I(\mathbf{x})\, \mathbf{a}_I\| \rightarrow 0 \quad \text{as} \quad h \rightarrow 0. \tag{4.55}$$

The condition can also be formulated for the entire domain. A suitable norm is needed to estimate the error. Often, the first Hilbert norm is chosen in the FEM for the analysis of the mechanical differential equation. This norm considers not only the values of the function $\mathbf{u}$ in an integral sense, but also the first derivative

$$\|\mathbf{u}\|_1 = \left(\int_B \left[\sum_{i=1}^{3} (u_i)^2 + \sum_{i=1}^{3} \sum_{j=1}^{3} \left(\frac{\partial u_i}{\partial x_j} \right)^2 \right] \mathrm{d}v \right)^{1/2}. \tag{4.56}$$

An overview of common norms in the context of numerical solution schemes can be found e.g. in Bathe (2006). A stricter criterion represents the reproducing condition (Belytschko et al. 2000b). In this case, it is required that the approximation by means of basis or shape functions corresponds exactly to the given function. Hence, in the neighborhood of $\mathbf{x}$ it has to hold

$$\mathbf{u}(\mathbf{x}) = \sum_{I=1}^{N_\mathbf{x}} N_I(\mathbf{x})\, \mathbf{u}(\mathbf{x}_I). \tag{4.57}$$

Compared to (4.1), the coefficients are replaced with the values of the function at the nodes or particles, i.e. $\mathbf{a}_I = \mathbf{u}(\mathbf{x}_I)$. As can be seen from the comparison with (4.55), completeness is always satisfied if the approximation fulfills the reproducing conditions.

Stability. In mechanics this term is often found in connection with equilibrium states. However, the solution procedure must also be stable. In general, stability requires boundedness. A system is stable if the values at two neighboring points $\mathbf{a}_a$ and $\mathbf{a}_b$ remain in the neighborhood during a process (Seydel 2009). The second term $\mathbf{a}_b$ can be considered as a small perturbation of $\mathbf{a}_a$. The distance, defined in an appropriate norm, must always be below a defined bound ϵ

$$\|\mathbf{a}_a(t) - \mathbf{a}_b(t)\| \leq \epsilon. \tag{4.58}$$

In the context of solution schemes, it is required that a small relative change of the load must lead to only a small relative change of the solution at any time (Bathe 2006). Therefore, the bound on the right-hand side in (4.58) is supplemented by the corresponding distances

$$\frac{\|\mathbf{a}_a(t) - \mathbf{a}_b(t)\|}{\|\mathbf{a}_a(t)\|} \leq \epsilon \frac{\|\mathbf{p}_a(t) - \mathbf{p}_b(t)\|}{\|\mathbf{p}_a(t)\|}. \tag{4.59}$$

Convergence. Convergence exists if the approximate $\mathbf{u}_h$ tends to the true solution $\mathbf{u}$ as the resolution increases

$$\|\mathbf{u}_h(\mathbf{x}) - \mathbf{u}(\mathbf{x})\| \rightarrow 0 \quad \text{as} \quad h \rightarrow 0. \tag{4.60}$$

The quantity h describing the resolution must be defined in an appropriate form. According to Lax's equivalence theorem in the context of the Finite Difference Method, the solution converges for a correctly posed problem if consistency and stability are ensured. No direct proof exists for other solution schemes. In the Finite Element Method using the isoparametric concept, convergence can be assumed for elliptic differential equations if the conditions on completeness and stability are satisfied.

In addition, the order of convergence α can be estimated with the help of an inequality, see e.g. Hughes (1987)

$$\|\mathbf{u}_h(\mathbf{x}) - \mathbf{u}(\mathbf{x})\|_s \leq c\, h^{\alpha} \|\mathbf{u}\|_r, \quad \alpha = \min(k + 1 - s, r - s). \tag{4.61}$$

If the solution can be assumed to be smooth, i.e., no drastic changes or jumps exist, r tends to infinity (Belytschko et al. 2000b). The value k denotes the order of completeness and s the order of the Hilbert norm. The value c depends on the problem and the material parameters but is independent of h.

For meshfree methods, two different types of convergence can be defined. A finer resolution always describes a larger number of nodes or particles. In convergence analysis, either the search radius can be assumed to be constant, or the number of nodes or particles in the neighborhood. The latter corresponds to the classical definition of convergence in the context of the FEM.

Patch test. In order to avoid a direct mathematical proof of consistency, the so-called patch test was introduced in Bazeley et al. (1965) in the context of the FEM. The justification of the equivalence was confirmed in Strang (1972), see also Strang and Fix (1973). The primary motivation was to investigate nonconforming elements. However, this test proved to be a very pragmatic approach to evaluate general spatial discretization concepts. The original form includes two test scenarios. If the values **a** of the true solution are substituted into the system of linear equations, see (4.37), the error must tend to zero as the resolution increases

$$\|\mathbf{Ka} - \mathbf{p}\| \rightarrow 0 \quad \text{as} \quad h \rightarrow 0. \tag{4.62}$$

In the second investigation, the values of the true solution are given only at the boundary. At the locations in the interior, the values **a** must correspond to the true solution. An extension of the original patch test can be found in Taylor et al. (1986). A third criterion is introduced which also tests the stability of the formulation. Thus, the patch test including all three scenarios can be regarded as a necessary and sufficient proof of convergence at least for the standard FEM using the isoparametric concept. In the additional test, Neumann boundary conditions are given. Only the minimum number of displacements is fixed to avoid rigid body motion. Convergence exists if the values **a** tend towards the true solution at a finer resolution.

Locking. This technical term is often mentioned in the context of the FEM. Locking expresses that the displacements from a simulation are smaller than the true values (Braess 2007). Belytschko et al. (2000b) suggests an addition. Locking further implies that the solution does not converge. If the solution converges only at a very fine resolution, the system is characterized as excessively stiff.

4.5.2 Reproducing Conditions

A central requirement for the trial function or approximate solution comprises the fulfillment of the completeness or more precisely the reproducing conditions. The weak form of the mechanical differential equation has order one, so the fulfillment of the reproducing conditions up to the 1st order is sufficient. The requirement is limited to polynomials. An illustrative justification can be found in Hughes (1987). Assuming that the number of points in the neighborhood remains constant, with a finer and finer resolution of the domain, the true solution and also its derivative approaches more and more a constant value in the neighborhoods. Therefore, the chosen approximation must be able to reproduce the constant and linear term of a polynomial function $\mathbf{p}(\mathbf{x})$. Based on (4.57) it has to hold

$$\sum_{I=1}^{N_{\mathbf{x}}} N_I(\mathbf{x})\, \mathbf{p}(\mathbf{x}_I) = \mathbf{p}(\mathbf{x}) \,. \tag{4.63}$$

Similarly, for the derivative it follows from (4.63)

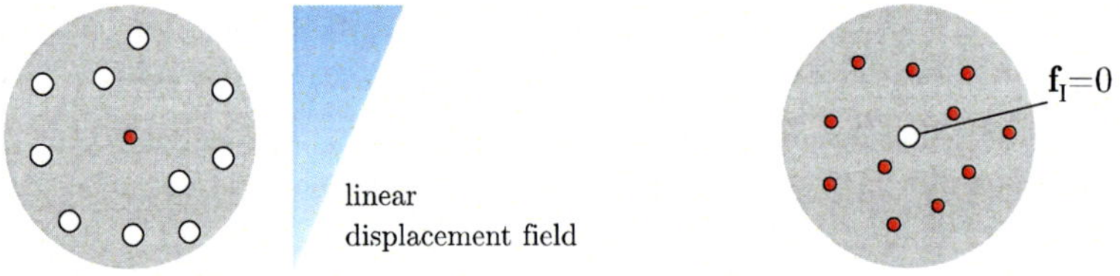

Fig. 4.5 (Left) 1st order reproducing condition in the derivatives of trial functions. (Right) Integration constraint for inner nodes

$$\sum_{I=1}^{N_{\mathbf{x}}} \mathbf{p}(\mathbf{x}_I) \otimes \frac{\partial N_I(\mathbf{x})}{\partial \mathbf{x}} = \frac{\partial \mathbf{p}(\mathbf{x})}{\partial \mathbf{x}}. \tag{4.64}$$

The 0th order condition is fulfilled if the constant term of the polynomial function can be reproduced, i.e., using $p(\mathbf{x}_I) = 1$ and $p(\mathbf{x}) = 1$ in (4.63) and (4.64)

$$\sum_{I=1}^{N_{\mathbf{x}}} N_I(\mathbf{x}) = 1, \qquad \sum_{I=1}^{N_{\mathbf{x}}} \frac{\partial N_I(\mathbf{x})}{\partial \mathbf{x}} = 0. \tag{4.65}$$

The extension to the condition of 1st order results from the consideration of the linear term, i.e. $\mathbf{p}(\mathbf{x}_I) = \mathbf{x}_I$ and $\mathbf{p}(\mathbf{x}) = \mathbf{x}$

$$\sum_{I=1}^{N_{\mathbf{x}}} N_I(\mathbf{x})\,\mathbf{x}_I = \mathbf{x}, \qquad \sum_{I=1}^{N_{\mathbf{x}}} \mathbf{x}_I \otimes \frac{\partial N_I(\mathbf{x})}{\partial \mathbf{x}} = \mathbf{1}. \tag{4.66}$$

A meaning of the reproduction condition of first order in the derivatives can be found in Fig. 4.5. If the strong form is solved, the conditions up to the 2nd order must be fulfilled. Hence, the shape function must also correctly reproduce the quadratic term with $\mathbf{p}(\mathbf{x}_I) = \mathbf{x}_I \otimes \mathbf{x}_I$ and $\mathbf{p}(\mathbf{x}) = \mathbf{x} \otimes \mathbf{x}$

$$\sum_{I=1}^{N_{\mathbf{x}}} N_I(\mathbf{x})\,\mathbf{x}_I \otimes \mathbf{x}_I = \mathbf{x} \otimes \mathbf{x}, \qquad \sum_{I=1}^{N_{\mathbf{x}}} \mathbf{x}_I \otimes \mathbf{x}_I \otimes \frac{\partial N_I(\mathbf{x})}{\partial \mathbf{x}} = \frac{\partial (\mathbf{x} \otimes \mathbf{x})}{\partial \mathbf{x}}. \tag{4.67}$$

In addition, the requirement for the second derivative must be satisfied

$$\sum_{I=1}^{N_{\mathbf{x}}} \mathbf{p}(\mathbf{x}_I) \otimes \frac{\partial^2 N_I(\mathbf{x})}{\partial^2 \mathbf{x}} = \frac{\partial^2 \mathbf{p}(\mathbf{x})}{\partial^2 \mathbf{x}}. \tag{4.68}$$

The reproducing conditions can alternatively be derived from a Taylor series. Such investigations are found in the context of Smoothed Particle Hydrodynamics, see e.g.. Bonet and Kulasegaram (2000) or Price (2012). The value at the particle or node can also be represented as a infinitesimal sum of derivatives at $\mathbf{x}$

$$\mathbf{p}(\mathbf{x}_I) = \mathbf{p}(\mathbf{x}) + \frac{\partial \mathbf{p}(\mathbf{x})}{\partial \mathbf{x}}(\mathbf{x}_I - \mathbf{x}) + \frac{1}{2}\frac{\partial^2 \mathbf{p}(\mathbf{x})}{\partial^2 \mathbf{x}}(\mathbf{x}_I - \mathbf{x})^2 + \mathcal{O}\left((\mathbf{x}_I - \mathbf{x})^3\right). \quad (4.69)$$

If the accuracy up to the 1st order is investigated, the components of higher order terms in the Taylor series can be neglected. Substituting (4.69) into (4.63) yields to the relation

$$\mathbf{p}(\mathbf{x}) = \sum_{I=1}^{N_\mathbf{x}} N_I(\mathbf{x})\,\mathbf{p}(\mathbf{x}) + \frac{\partial \mathbf{p}(\mathbf{x})}{\partial \mathbf{x}} \sum_{I=1}^{N_\mathbf{x}} N_I(\mathbf{x})(\mathbf{x}_I - \mathbf{x}). \quad (4.70)$$

A 1st order accuracy in the approximation exists when the shape functions in the first term sum to one and the second term on the right-hand side vanishes

$$\sum_{I=1}^{N_\mathbf{x}} N_I(\mathbf{x}) = 1, \quad \sum_{I=1}^{N_\mathbf{x}} N_I(\mathbf{x})(\mathbf{x}_I - \mathbf{x}) = 0. \quad (4.71)$$

Comparing with (4.65) and (4.66), the requirements of 1st order accuracy in the approximation is equal to the reproducing conditions up to the 1st order. Similarly, conditions can be defined for the accuracy of 1st order in the derivatives by substituting (4.69) into (4.64)

$$\frac{\partial \mathbf{p}(\mathbf{x})}{\partial \mathbf{x}} = \mathbf{p}(\mathbf{x}) \otimes \sum_{I=1}^{N_\mathbf{x}} \frac{\partial N_I(\mathbf{x})}{\partial \mathbf{x}} + \frac{\partial \mathbf{p}(\mathbf{x})}{\partial \mathbf{x}} \sum_{I=1}^{N_\mathbf{x}} (\mathbf{x}_I - \mathbf{x}) \otimes \frac{\partial N_I(\mathbf{x})}{\partial \mathbf{x}}. \quad (4.72)$$

The equal sign applies only if the reproducing conditions up to 1st order in the derivatives (4.65), (4.66) are satisfied.

4.5.3 *Integration Constraint*

In the Finite Element Method, the fulfillment of the patch test together with a sufficiently accurate integration is considered as necessary criteria for convergence of elliptic differential equations. If the elements are conforming, see Sect. 4.5.6, and the shape functions fulfill the required reproducing conditions, the patch test is usually satisfied (Hughes 1987). For spatial discretization schemes based on the weak form a precise criterion can be formulated which is based on the investigations in Krongauz and Belytschko (1997). Due to the connection with the test function and the integration, the name integration constraint was introduced in Chen et al. (2001). A generalization can be found in Chen et al. (2013). Therein, the name was further changed to variational consistency. The criterion is based on the requirement that the spatial discretization scheme must satisfy Gauss's integral theorem

$$\int_{\partial B} \boldsymbol{\eta} \cdot \boldsymbol{\sigma} \mathbf{n} \, da - \int_B \operatorname{grad} \boldsymbol{\eta} \cdot \boldsymbol{\sigma} \, dv = \int_B \boldsymbol{\eta} \cdot \operatorname{div} \boldsymbol{\sigma} \, dv. \tag{4.73}$$

In the discrete form, a condition on the test shape function and on the integration can be obtained for each node or particle. The evaluation of the integral can be limited to the neighborhood H_I of each particle I, since the shape function is equal to zero outside H_I

$$\int_{\partial H_I} N_I(\mathbf{x}) \boldsymbol{\sigma} \mathbf{n} \, da - \int_{H_I} \boldsymbol{\sigma} \frac{\partial N_I(\mathbf{x})}{\partial \mathbf{x}} \, dv = \int_{H_I} N_I(\mathbf{x}) \operatorname{div} \boldsymbol{\sigma} \, dv. \tag{4.74}$$

From this general context, different orders of the integration constraint can be derived. According to the Weierstrass theorem, the stress can be represented by variables, summarized in the vector $\mathbf{p}(\mathbf{x})$, and suitable coefficients $\mathbf{d}$

$$\boldsymbol{\sigma} = \mathbf{p}(\mathbf{x}) \mathbf{d}. \tag{4.75}$$

The coefficients can take on any non-zero values. Substituting (4.75) into (4.74) yields to the final criterion

$$\int_{\partial H_I} N_I(\mathbf{x}) \mathbf{p}(\mathbf{x}) \otimes \mathbf{n} \, da - \int_{H_I} \mathbf{p}(\mathbf{x}) \otimes \frac{\partial N_I(\mathbf{x})}{\partial \mathbf{x}} \, dv = \int_{H_I} N_I(\mathbf{x}) \frac{\partial \mathbf{p}(\mathbf{x})}{\partial \mathbf{x}} \, dv. \tag{4.76}$$

Linear displacements lead to constant stresses. The discretization scheme is therefore considered to be linearly exact if, in addition to the reproducing condition of 1st order, (4.74) is satisfied for constant stresses

$$\int_{\partial H_I} N_I(\mathbf{x}) \mathbf{n} \, da - \int_{H_I} \frac{\partial N_I(\mathbf{x})}{\partial \mathbf{x}} \, dv = 0. \tag{4.77}$$

This requirement also follows from (4.76) with $\mathbf{p}(\mathbf{x}) = 1$. If the displacements are quadratically distributed then

$$\int_{\partial H_I} N_I(\mathbf{x}) \mathbf{x} \otimes \mathbf{n} \, da - \int_{H_I} \mathbf{x} \otimes \frac{\partial N_I(\mathbf{x})}{\partial \mathbf{x}} \, dv = \int_{H_I} N_I(\mathbf{x}) \mathbf{1} \, dv \tag{4.78}$$

must hold for the discretization scheme to be quadratically exact. Equation (4.78) in turn follows from (4.76) with $\mathbf{p}(\mathbf{x}) = \mathbf{x}$. For the solution of the mechanical differential equation in the weak form, linear exactness is sufficient to pass the standard patch test.

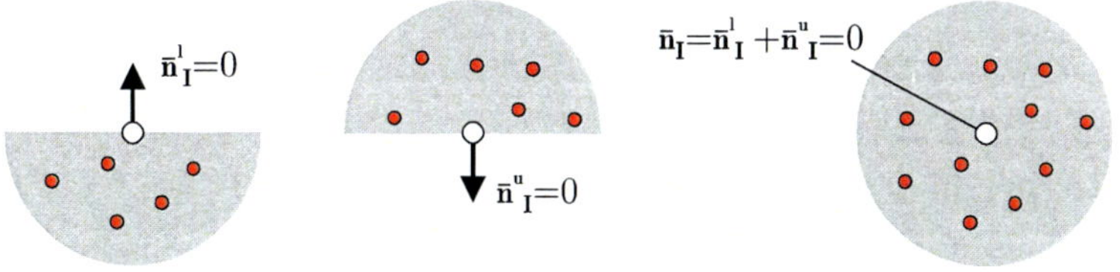

Fig. 4.6 Cut through a support domain showing the resulting normal vectors at the boundary nodes due to the integration constraint

When using numerical integration, (4.77) results in the following requirement in case of a two-point method

$$\sum_{p=1}^{N_I^{inf}} \frac{\partial N_I(\mathbf{x}_p)}{\partial \mathbf{x}_p} v_p = \begin{cases} 0 & \text{inner nodes} \\ \bar{\mathbf{n}}_I & \text{boundary nodes} \end{cases} . \tag{4.79}$$

This condition is equivalent to the requirement that at a constant stress the force at inner nodes must be zero (Fig. 4.5). The resulting normal vector $\bar{\mathbf{n}}_I$ is calculated from the position vector $\mathbf{x}_p^s$, the normalized normal vector $\mathbf{n}_p$, and the area a_p at the integration points

$$\bar{\mathbf{n}}_I = \sum_{p=1}^{N_I^{s,inf}} N_I(\mathbf{x}_p^s)\mathbf{n}_p a_p. \tag{4.80}$$

The integral is replaced by the sum over all integration points at the surface $N_I^{s,inf}$ associated with the boundary node I.

The vector $\bar{\mathbf{n}}_I$ can also be interpreted as a sum over normalized normal vectors $\mathbf{n}_{I\,p}$ multiplied by the corresponding area fractions $a_{I\,p}$ (Weißenfels 2019)

$$\bar{\mathbf{n}}_I = \sum_{p=1}^{N_I^{inf}} \mathbf{n}_{I\,p} a_{I\,p} = \sum_{p=1}^{N_I^{inf}} \bar{\mathbf{n}}_{I\,p} = \sum_{p=1}^{N_I^{inf}} \frac{\partial N_I(\mathbf{x}_p)}{\partial \mathbf{x}_p} v_p. \tag{4.81}$$

For interior nodes, the integration constraint is thus equivalent to the requirement that the sum of all normal vectors of a node evaluated at the integration points $\bar{\mathbf{n}}_{I\,p}$ must be zero. This requirement also corresponds to Stokes theorem (Fleming 2012). At the boundary, the node is not completely surrounded by integration points and the surface integral is not closed. The missing part corresponds to the resulting normal vector at the surface, see Fig. 4.6.

The standard FEM always fulfills the integration constraint due to the isoparametric concept as demonstrated in Weißenfels (2019). In this case, only the nodes at the boundary have a resulting normal vector other than zero, see also Fig. 4.7. Hence, the nodes at the boundary can be detected directly without knowing the surface. If

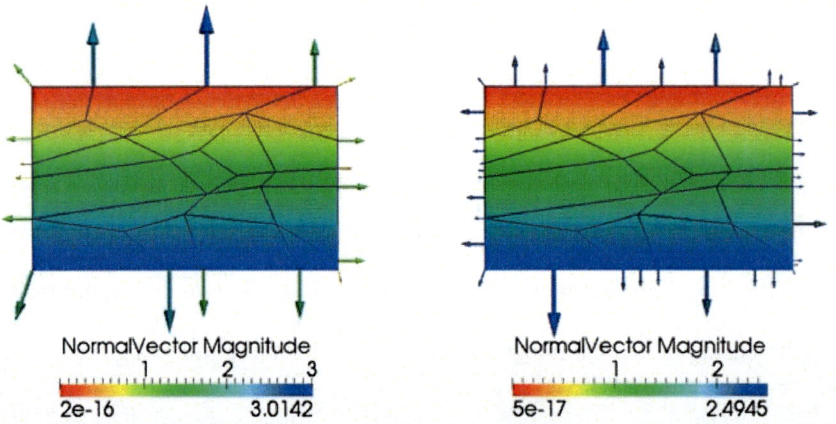

Fig. 4.7 Linear displacement field and resulting normal vectors for the uniaxial loading of a block discretized with irregular element. (Left) Linear polynomial basis function (Q1). (Right) Quadratic polynomial basis function (Q2)

the norm of the resulting normal vector is added up over all nodes, the entire surface is obtained.

Since the integration constraint provides not only the resulting normal vector but also the associated nodal area, surface loads can be directly converted into body forces as shown in Weißenfels (2019) for the FEM.

Remark 1. For meshfree methods, the continuous surface force (CSF) approach of Brackbill et al. (1992) is often used. The resulting normal vector is determined from a color function c

$$\bar{\mathbf{n}}(\mathbf{x}) = \operatorname{grad} c(\mathbf{x}). \tag{4.82}$$

The quantity c may correspond to the density (Brackbill et al. 1992) or the mass (Chen et al. 2012). A similar approach is found in the SPH to calculate the normal vector at the particle (Randles and Libersky 1996). However, these forms do not correspond to the integration constraint, since the equations are not derived from Gauss's integral theorem.

Remark 2. In Bubnov-Galerkin methods, the shape function must satisfy both the reproducing conditions and the integration constraint. For Petrov-Galerkin methods, on the other hand, the requirements can be considered separately, see also Sect. 5.2.6.

4.5.4 *Configurational Consistency*

The solution algorithm must ensure that the calculated quantities in the current configuration can also be determined via a push-forward from a previous configuration. This has to hold especially for the derivative of the shape function and for line elements. If the set of points in a neighborhood is constant during the calculation, quantities can also be determined with respect to the initial configuration. If the derivative of the shape function is calculated directly in the current configuration, then, e.g. for a two-point method, the following must be valid

$$\frac{\partial N_I(\mathbf{x}_{p\,n+1})}{\partial \mathbf{x}_{n+1}} = \frac{\partial N_I(\mathbf{x}_{p\,n})}{\partial \mathbf{x}_n} \Delta \mathbf{F}^{-1}_{p\,n+1} = \frac{\partial N_I(\mathbf{X})}{\partial \mathbf{X}} \mathbf{F}^{-1}_{p\,n+1}. \tag{4.83}$$

The same applies for the update of line elements. The coordinates calculated by the solution algorithm must also be determined by the deformation gradient and the corresponding coordinates in the initial or previous configuration

$$\mathbf{x}_{I\,n+1} - \mathbf{x}_{p\,n+1} = \Delta \mathbf{F}_{p\,n+1} \left(\mathbf{x}_{I\,n} - \mathbf{x}_{p\,n} \right) = \mathbf{F}_{p\,n+1} \left(\mathbf{X}_I - \mathbf{X}_p \right). \tag{4.84}$$

4.5.5 Discrete Conservation Properties

Discretization schemes must not violate any balance laws. In a purely mechanical view, the conservation of momentum and angular momentum must be satisfied in the discrete case as well.

Global balance of momentum. This law postulates that the change in momentum **L** over time is equal to the external forces. If no external forces are acting, the change of momentum must be equal to zero

$$\frac{\mathrm{d}}{\mathrm{d}t} \mathbf{L} = \sum_{I=1}^{n} \mathbf{f}_I^t = 0. \tag{4.85}$$

The abbreviation n corresponds to the total number of nodes or particles in the domain. Conversely, it follows from Eqs. (4.2) and (4.3) that the sum of all internal forces due to the state of stress in the material must be zero

$$\sum_{I=1}^{n} \mathbf{f}_I^t = -\sum_{I=1}^{n} \mathbf{f}_I^s = 0. \tag{4.86}$$

Substituting the corresponding $\mathbf{f}_I^s$ from Sect. 4.2, concrete requirements can be stated for each concept, like for the two-point method based on the weak form. By substituting (4.9) into (4.86) and swapping the order of summation, the requirement can be specified more precisely

$$\sum_{I=1}^{n} \mathbf{f}_I^s = \sum_{I=1}^{n} \sum_{p=1}^{N_I^{inf}} \boldsymbol{\sigma}_p \frac{\partial N_I(\mathbf{x}_p)}{\partial \mathbf{x}_p} v_p = \sum_{p=1}^{n_{gp}} \boldsymbol{\sigma}_p \, v_p \sum_{I=1}^{N_p^{sup}} \frac{\partial N_I(\mathbf{x}_p)}{\partial \mathbf{x}_p} = 0. \tag{4.87}$$

Hence, in order not to violate the conservation of momentum, the test shape function must also satisfy the reproducing condition of 0th order in the derivatives.

Global balance of angular momentum. The second condition requires that the angular momentum $\mathbf{J}$ is conserved. If again external forces are neglected, the change of $\mathbf{J}$ over time must be zero

$$\frac{\mathrm{d}}{\mathrm{d}t}\mathbf{J} = \sum_{I=1}^{n} \mathbf{f}_I^t \times \mathbf{x}_I = \sum_{I=1}^{n} \mathcal{E} : \mathbf{x}_I \otimes \mathbf{f}_I^t = 0. \tag{4.88}$$

With the help of the permutation tensor $\mathcal{E}$ the cross product can be reformulated. The operator : corresponds to the double contraction. A precise definition of the 3rd order tensor $\mathcal{E}$ can be found, for example, in Holzapfel (2000). Due to (4.86) the requirement can also be related to the internal force

$$-\sum_{I=1}^{n} \mathcal{E} : \mathbf{x}_I \otimes \mathbf{f}_I^s = 0. \tag{4.89}$$

Substituting $\mathbf{f}_I^s$ from Sect. 4.2 the criterion can be specified. When using a two-point method based on the weak form, the requirement can be formulated more precisely by swapping the sum signs and exploiting the symmetry of the stress tensor

$$\sum_{I=1}^{n} \mathcal{E} : \mathbf{x}_I \otimes \sum_{p=1}^{N_I^{int}} \boldsymbol{\sigma}_p \frac{\partial N_I(\mathbf{x}_p)}{\partial \mathbf{x}_p} v_p = \sum_{p=1}^{n_{gp}} \mathcal{E} : \left(\sum_{I=1}^{N_p^{sup}} \mathbf{x}_I \otimes \frac{\partial N_I(\mathbf{x}_p)}{\partial \mathbf{x}_p} \right) \boldsymbol{\sigma}_p v_p = 0. \tag{4.90}$$

Since $\boldsymbol{\sigma}$ is symmetric, the relation $\mathcal{E} : \boldsymbol{\sigma} \equiv 0$ holds. Thus, the above equation is always satisfied if the test shape function fulfills the reproducing condition of 1st order in the derivatives. A similar conclusion is drawn in Belytschko et al. (1998).

4.5.6 *Continuity*

The condition of continuity refers to the shape functions within and across the neighborhoods. It is required that the derivatives are continuous up to the order p of the differential equation (C^p continuity). Across the boundary C^{p-1} continuity is sufficient. These two requirements ensure that the integration is well-defined (Hughes 1987). In the FEM, one speaks of conforming elements if continuity is given.

4.5.7 Kronecker-δ Property

An advantageous but not necessary requirement is the interpolation property of the shape function. This is given if the coefficient $\mathbf{a}_I$ is equal to the value of the function at the corresponding node or particle

$$\mathbf{u}(\mathbf{x}_I) = \sum_{J=1}^{N_\mathbf{x}} N_J(\mathbf{x}_I)\mathbf{a}_J = \mathbf{a}_I. \tag{4.91}$$

The above equation is satisfied only if the shape function possesses the Kronecker-δ property

$$N_J(\mathbf{x}_I) = \delta_{IJ}. \tag{4.92}$$

If (4.91) holds, Dirichlet boundary conditions can be applied directly. Otherwise, the imposition of these conditions is a constraint and special schemes, like the Lagrange multiplier or the penalty method, are needed.

4.5.8 Search Algorithm

The first step in discretization schemes is the determination of the neighborhoods. With meshfree methods, this assignment follows from a search algorithm. In the 3-dimensional case, the search area is often spherical, cuboidal, or elliptical. For a spherical region, the point J is part of the set H_P of point P if the distance $\|\mathbf{x}_J - \mathbf{x}_P\|$ is smaller than a certain radius R

$$H_P = \left\{\mathbf{x}_J \in B_t \middle| \|\mathbf{x}_J - \mathbf{x}_P\| < R\right\}. \tag{4.93}$$

The search algorithm itself must also meet several requirements. The distribution of the points must allow the shape function to be computed. For moving least square functions this results in a condition on the invertibility of the moment matrix and for local maximum entropy functions on the convergence of the Newton Raphson algorithm. In Liu et al. (1997), see also Li and Liu (2007), some requirements are formulated to ensure a stable and accurate calculation. Therefore, the distribution of the points has to be isotropic and homogeneous in the statistical mean to avoid degenerate states. In the 1-dimensional case, it must be ensured that at least two points form a line. In the 2-dimensional case, the requirement expands to 3 points forming a triangle and in the 3-dimensional continuum, there must be four points spanning a tetrahedron.

Furthermore, the search algorithm must not have any influence on the equilibrium of forces. If the neighborhood of point I is newly determined the force before and after the recomputation of the neighborhood must be identical

$$\mathbf{f}_{I\,n} = \mathbf{f}^0_{I\,n+1}. \tag{4.94}$$

Otherwise, unphysical forces occur, leading to inaccurate solutions. The index 0 denotes the force before the calculation at the considered time or load step. An investigation on this condition can be found in Sect. 5.2.2.

Formulations in SPH or Peridynamics that are based on forces between two points have to meet another requirement. The search algorithm must ensure that the global balance of momentum and angular momentum are not violated. A detailed description of this requirement can be found in Chaps. 6 and 7.

4.5.9 Stability

Spatial discretization schemes are prone to two different types of instabilities that must be avoided to get a good approximation of the solution. The first results from a poor evaluation of the discrete equations and the second type can be observed particularly clearly with loadings that lead to tensile stresses.

Rank instability. In the Finite Element Method, under-integration leads to a rank deficiency of the stiffness matrix. The spectral decomposition of the element tangent $\mathbf{K}^e$

$$\det\left(\mathbf{K}^e - \omega\mathbf{1}\right) = 0 \tag{4.95}$$

into eigenvalues ω, and eigenvectors, the so-called modes, allows the reasons to be identified more precisely. The eigenvalues belonging to the rigid body modes are always equal to zero. Under-integration leads to the fact that also the eigenvalues which belong to other modes are too small. For example, with a 2-dimensional four-node element and a 1-point integration, the eigenvalues of the bending modes are equal to zero. Thus, no strain energy can be assigned to these deformations, see also, e.g., Wriggers (2008). Conversely, many deformation states lead to the same strain energy. If these zero or low energy modes are excited by the external load, an unphysical solution of the differential equation results.

The same phenomena occur with meshfree methods, since there, too, the equations are evaluated only at discrete locations in the respective neighborhood. By selecting rational or exponential functions, it is difficult to determine in advance the number of points for a sufficiently accurate evaluation. Analogous to the FEM, it can be shown that no energies can be assigned to different deformation states if a sufficient number of evaluation points is not available (Tupek and Radovitzky 2014). For one-point methods, the possibilities to counteract under-integration are limited, since the equations can be evaluated only at the nodes or particles. Even with meshfree methods, the spectral decomposition of the stiffness matrix is a suitable approach to detect a rank instability, see e.g. Puso et al. (2008) or Chen et al. (2002).

Tensile instability. In meshfree methods, another instability can be detected, which is especially visible if the loading leads to tensile stresses. According to the study in Belytschko et al. (2000a), this instability occurs only when the equations are formulated with respect to the deformed (current, updated Lagrangian) configuration. A 1-dimensional analysis in Swegle et al. (1995) provides an intuitive explanation for this phenomenon. It is demonstrated that the stresses inside the body decrease even though the bar lengthens. This situation occurs when the second derivative of the weighting function is positive. Similar unphysical effects could also be proven for compressive stresses. In Swegle et al. (1995) it is also shown that an SPH formulation with respect to the deformed configuration is unstable when the product of the second derivative of the weighting function with the stress is larger than zero.

4.6 Shape Functions

In the FEM, the trial and test function is usually approximated with the aid of polynomial functions defined on a fixed, local reference element. The connection to the global quantities results from a coordinate transformation. With meshfree methods, the number of points in the neighborhood can change continuously. For this reason, shape functions are needed, which can approximate the solution function with an arbitrarily large number and an arbitrarily distributed position of points in the neighborhood. Over the years, different functions have been developed. As an example, four frequently used variants are presented in more detail.

4.6.1 *Least Square Functions*

According to Weierstrass' theorem, functions in a bounded and real interval can be approximated sufficiently accurately via a polynomial

$$\mathbf{u}\,(\mathbf{x}) = \mathbf{p}\,(\mathbf{x})\,\mathbf{d}. \tag{4.96}$$

The tensor $\mathbf{d}$ contains all unknown coefficients and the base function vector $\mathbf{p}\,(\mathbf{x})$ contains the variables of the polynomial

$$\mathrm{p}_i = 1, x_1, x_2, x_3, x_1^2, x_1 x_2, x_1 x_3, \ldots \tag{4.97}$$

From (4.96), shape functions can be derived if the coefficients $\mathbf{d}$ can be related to the coefficients $\mathbf{a}$ of the approximate solution (4.1). This connection can be constructed from the interpolation requirement. In this case, the function values at the node or particle are equal to the coefficients. Substituting (4.96) into (4.91) yields to the relationship between $\mathbf{a}$ and $\mathbf{d}$

$$\mathbf{a}_J = \mathbf{u}\,(\mathbf{x}_J) = \mathbf{p}\,(\mathbf{x}_J)\,\mathbf{d}. \tag{4.98}$$

With an arbitrary number of points in the neighborhood $N_{\mathbf{x}}$, an interpolation is difficult to implement. Using the method of least squares, the error in the approximation can be kept small

$$J(\mathbf{d}) = \sum_{J=1}^{N_{\mathbf{x}}} \|\mathbf{a}_J - \mathbf{p}(\mathbf{x}_J)\,\mathbf{d}\|^2 \;\rightarrow\; \text{min.} \tag{4.99}$$

For the functional J to be minimal the derivative with respect to $\mathbf{d}$ must be equal to zero. From this condition, the unknown coefficients $\mathbf{d}$ can be determined

$$\frac{\partial J}{\partial \mathbf{d}} = 0 \rightarrow \mathbf{d} = \mathbf{M}^{-1} \sum_{J=1}^{N_{\mathbf{x}}} \mathbf{p}(\mathbf{x}_J) \otimes \mathbf{a}_J, \quad \mathbf{M} = \sum_{J=1}^{N_{\mathbf{x}}} \mathbf{p}(\mathbf{x}_J) \otimes \mathbf{p}(\mathbf{x}_J). \tag{4.100}$$

The moment tensor $\mathbf{M}$ defines a complete dyadic product. Hence the inverse can exist, see also remark 2. Substituting (4.100) into (4.96) yields to the relation between $\mathbf{u}$ and $\mathbf{a}$ for an arbitrary number of points in the neighborhood

$$\mathbf{u}(\mathbf{x}) = \mathbf{p}(\mathbf{x}) \cdot \mathbf{M}^{-1} \sum_{J=1}^{N_{\mathbf{x}}} \mathbf{p}(\mathbf{x}_J)\,\mathbf{a}_J. \tag{4.101}$$

The first three factors together yield a scalar that can be identified as a shape function

$$\mathbf{u}(\mathbf{x}) = \sum_{J=1}^{N_{\mathbf{x}}} N_J(\mathbf{x})\,\mathbf{a}_J, \quad N_J(\mathbf{x}) = \mathbf{p}(\mathbf{x}) \cdot \mathbf{M}^{-1}\mathbf{p}(\mathbf{x}_J). \tag{4.102}$$

The above relation does not correspond to an interpolation, since $\mathbf{a}_J \neq \mathbf{u}(\mathbf{x}_J)$. The error can be estimated from (4.99). The drawback of the classical method of least squares is the increasing degradation of the approximation when the number of particles or nodes significantly exceeds the number of entries in $\mathbf{p}$ (Onate et al. 1996). A local weighting of the error at a certain point $\mathbf{x}_I$ leads to a better approximation in this case

$$J(\mathbf{d}_I) = \sum_{J=1}^{N_{\mathbf{x}}} \omega(\mathbf{x}_J - \mathbf{x}_I)\left[\mathbf{a}_J - \mathbf{p}(\mathbf{x}_J)\,\mathbf{d}_I\right] \rightarrow \text{min.} \tag{4.103}$$

The weighting function ω usually corresponds to a window function that depends on the distance between $\mathbf{x}_J$ and $\mathbf{x}_I$. In the weighted least square approach, the unknown coefficients and the moment tensor are a function of $\mathbf{x}_I$

$$\mathbf{M}(\mathbf{x}_I) = \sum_{J=1}^{N_{\mathbf{x}}} \omega(\mathbf{x}_J - \mathbf{x}_I)\,\mathbf{p}(\mathbf{x}_J) \otimes \mathbf{p}(\mathbf{x}_J). \tag{4.104}$$

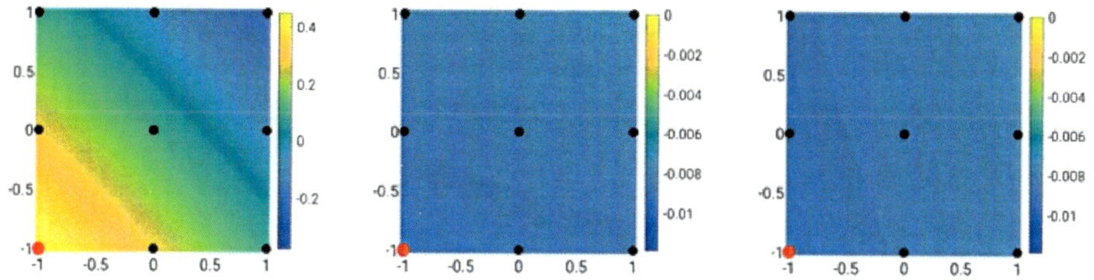

Fig. 4.8 (Left) Distribution of the moving least square (MLS) shape function of the lower left node using a linear polynomial basis function. (Middle) Derivative in x-direction. (Right) Derivative in y-direction

The same holds for the resulting shape function

$$N_J(\mathbf{x}, \mathbf{x}_I) = \omega(\mathbf{x}_J - \mathbf{x}_I)\,\mathbf{p}(\mathbf{x}) \cdot \mathbf{M}(\mathbf{x}_I)^{-1}\,\mathbf{p}(\mathbf{x}_J)\,. \tag{4.105}$$

Moving least square functions follow from the transition $\mathbf{x}_I \rightarrow \mathbf{x}$

$$\begin{aligned} N_J(\mathbf{x}) &= \omega(\mathbf{x}_J - \mathbf{x})\,\mathbf{p}(\mathbf{x}) \cdot \mathbf{M}(\mathbf{x})^{-1}\,\mathbf{p}(\mathbf{x}_J)\,, \\ \mathbf{M}(\mathbf{x}) &= \sum_{J=1}^{N_\mathbf{x}} \omega(\mathbf{x}_J - \mathbf{x})\,\mathbf{p}(\mathbf{x}_J) \otimes \mathbf{p}(\mathbf{x}_J)\,. \end{aligned} \tag{4.106}$$

Least square shape functions fulfill the reproducing conditions up to the chosen degree of the polynomial, but do not possess the Kronecker-δ property. This can be seen in Fig. 4.8 which shows moving least square functions based on

$$w(\mathbf{x}_J - \mathbf{x}) = \begin{cases} \frac{2}{3} - 4r^2 + 4r^3 & 0 \le r \le \frac{1}{2} \\ \frac{4}{3} - 4r + 4r^2 - \frac{4}{3}r^3 & \frac{1}{2} < r \le 1 \\ 0 & r > 1 \end{cases}, \qquad r = \frac{\|\mathbf{x}_J - \mathbf{x}\|}{d}. \tag{4.107}$$

An introduction into approximations using least squares presenting more details can be found in Lancaster and Salkauskas (1986).

Remark 1. A window function results in a scalar and is often used in signal processing. The values are non-zero only in an interval. Mostly the function is symmetrical around the maximum, which is in the middle and tapers towards the boundary of the interval.

Remark 2. Tensor or dyadic products, respectively, can be assigned to different classes, depending on the rank of the tensor (Lindell 1992). A complete dyadic product exists if the matrix possesses a full rank. Thus, the determinant is also nonzero.

Remark 3. With the choice of a constant polynomial function $\mathbf{p}(\mathbf{x}) = 1$ (4.106) reduces to the Shepard function

$$N_J(\mathbf{x}) = \frac{w(\mathbf{x}_J - \mathbf{x})}{W}, \quad W = \sum_{J=1}^{N_x} w(\mathbf{x}_J - \mathbf{x}). \tag{4.108}$$

4.6.2 Radial Functions

Alternatives to polynomials are radial functions. This type is based on the distance between points. The reproducing conditions are often not met and suitable adaptations are needed. Exemplary the C^2 function of the Wendland class (Wendland 1995) is presented which is often used as a weighting or kernel function in the SPH. The 0th order reproducing condition can be satisfied by a normalization

$$N_I(\mathbf{x}) = \frac{w_I(\mathbf{x})}{W}, \quad w_I(\mathbf{x}) = (1 + \frac{3}{2}r) * (2 - r)^3, \quad W = \sum_{I=1}^{N} w_I(\mathbf{x}). \tag{4.109}$$

The definition of the normalized distance r is given in (4.107). In the context of SPH mostly only the derivative of the function is corrected to fulfill the 1st order reproducing conditions in the derivatives. Since the standard function violates this criterion a tensor $\mathbf{L}$ results from (4.66) instead of the unit tensor

$$\mathbf{L}(\mathbf{x}) = \sum_{I=1}^{N_x} \mathbf{x}_I \otimes \frac{\partial N_I(\mathbf{x})}{\partial \mathbf{x}}. \tag{4.110}$$

This tensor can now be used to restore the 1st order reproducing condition in the derivatives

$$\frac{\partial N_I^c(\mathbf{x})}{\partial \mathbf{x}} = \frac{\partial N_I(\mathbf{x})}{\partial \mathbf{x}} \mathbf{L}^{-1}, \quad \text{leading to} \quad \sum_{I=1}^{N_x} \mathbf{x}_I \otimes \frac{\partial N_I^c(\mathbf{x})}{\partial \mathbf{x}} = \mathbf{1}. \tag{4.111}$$

This approach goes back to Randles and Libersky (1996). In addition, other correction methods exist. Some of them are presented in Sect. 6.5.

The corrected Wendland C^2 functions do not possess the Kronecker-δ property, as can be seen in Fig. 4.9.

4.6.3 Local Maximum Entropy Functions

The local maximum entropy (LME) function, presented in Arroyo and Ortiz (2006), is motivated by information theory and statistical mechanics. An introduction into these topics can be found in Jaynes (1957)

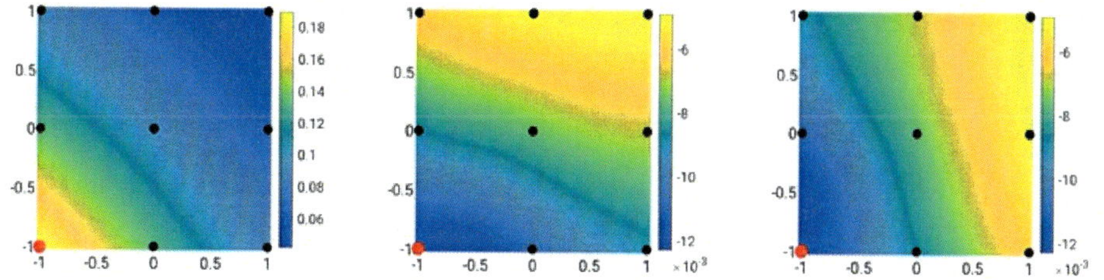

Fig. 4.9 (Left) Distribution of the corrected 1-dimensional Wendland C^2 shape function of the lower left node. (Middle) Derivative in x-direction. (Right) Derivative in y-direction

$$N_I\left(\mathbf{x},\boldsymbol{\lambda}^*\right)=\frac{1}{Z\left(\mathbf{x},\boldsymbol{\lambda}^*\right)}\exp\bigl(-\|\mathbf{x}-\mathbf{x}_I\|_\beta^2+\boldsymbol{\lambda}^*\cdot(\mathbf{x}-\mathbf{x}_I)\bigr). \tag{4.112}$$

The 1st order reproducing condition is defined as a constraint, which is enforced by the Lagrange multiplier method. The superscript * indicates that the Lagrange multiplier $\boldsymbol{\lambda}$ must be determined at each evaluation point $\mathbf{x}$ separately. Different variants exist for the first term in the exponent. The original form weights only the square distance with a freely selectable parameter β (Arroyo and Ortiz 2006). In Kumar et al. (2019) a more general form is found, in which a parameter matrix $\boldsymbol{\beta}$ is multiplied on both sides by a difference

$$\|\mathbf{x}-\mathbf{x}_I\|_\beta^2=\begin{cases}\beta\|\mathbf{x}-\mathbf{x}_I\|^2\\(\mathbf{x}-\mathbf{x}_I)\;\boldsymbol{\beta}\cdot(\mathbf{x}-\mathbf{x}_I)\end{cases}. \tag{4.113}$$

The fulfillment of the reproducing condition of 0th order can be achieved by a normalization using

$$Z\left(\mathbf{x},\boldsymbol{\lambda}^*\right)=\sum_{I=1}^{N_\mathbf{x}}\exp\bigl(-\|\mathbf{x}-\mathbf{x}_I\|_\beta^2+\boldsymbol{\lambda}^*\cdot(\mathbf{x}-\mathbf{x}_I)\bigr). \tag{4.114}$$

Sought is the Lagrange multiplier, which ensures that the 1st order reproducing condition is satisfied

$$\mathbf{r}\left(\mathbf{x},\boldsymbol{\lambda}\right)=\sum_{I=1}^{N_\mathbf{x}}N_I\left(\mathbf{x},\boldsymbol{\lambda}\right)\left(\mathbf{x}-\mathbf{x}_I\right)\overset{!}{=}0. \tag{4.115}$$

To calculate the desired $\boldsymbol{\lambda}^*$, the Newton-Raphson algorithm from Sect. 4.3 can be applied. As pointed out in Foca (2015), the use of a regularized Newton method facilitate finding the solution in critical cases. The LME functions possess a weak Kronecker-δ property. Hence, the function is interpolative only at the boundary of a convex domain (Fig. 4.10). Only there Dirichlet conditions can be directly imposed.

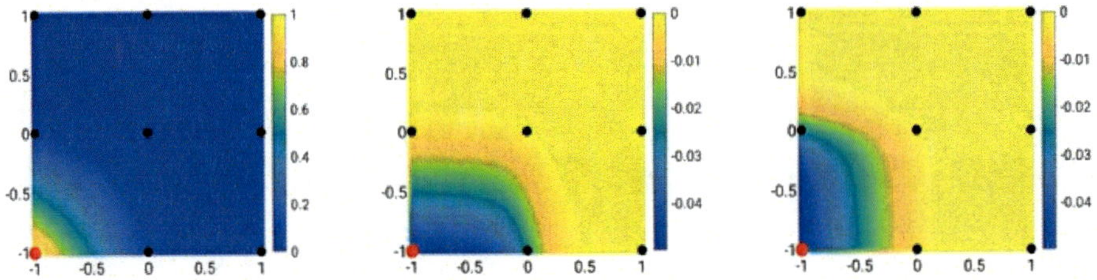

Fig. 4.10 (Left) Distribution of the local maximum entropy (LME) shape function of the lower left node. (Middle) Derivative in x-direction. (Right) Derivative in y-direction

4.6.4 Integral Functions

In Smoothed Particle Hydrodynamics, in the Reproducing Kernel Particle Method or in the Moving Least-Square Particle Method the shape functions are based on a linear integral operator T. With the help of a kernel function K a function $\mathbf{u}$ is mapped to another function via an integral over the neighborhood $H_{\mathbf{x}}$ of $\mathbf{x}$

$$(T\mathbf{u})(\mathbf{x}) = \int_{H_{\mathbf{x}}} K\left(\mathbf{x}, \mathbf{x}'\right) \mathbf{u}\left(\mathbf{x}'\right) \, \mathrm{d}v_{\mathbf{x}'}. \tag{4.116}$$

A convolution results if K depends on the difference between $\mathbf{x}$ and $\mathbf{x}'$

$$(T\mathbf{u})(\mathbf{x}) = (K * \mathbf{u})(\mathbf{x}) = \int_{H_{\mathbf{x}}} K\left(\mathbf{x} - \mathbf{x}'\right) \mathbf{u}\left(\mathbf{x}'\right) \, \mathrm{d}v_{\mathbf{x}'}. \tag{4.117}$$

In this case, K is also called weighting function. The integral operator has a reproducing property when K maps the function to itself (Aronszajn 1950)

$$(T\mathbf{u})(\mathbf{x}) = \mathbf{u}(\mathbf{x}) = \int_{H_{\mathbf{x}}} K\left(\mathbf{x}, \mathbf{x}'\right) \mathbf{u}\left(\mathbf{x}'\right) \, \mathrm{d}v_{\mathbf{x}'}. \tag{4.118}$$

An analogous relation can be stated for the convolution (4.117). In the literature, different approaches can be found to derive shape functions from integral transforms. Three variants are presented in more detail.

Variant 1. Similar to Sect. 4.6.1, the shape functions derived in Liu et al. (1995) are based on polynomials. However, the unknown coefficients $\mathbf{d}$ from (4.96) are calculated from (4.117) and not via the method of least square. Instead of K, a window function ω is used. In the version of Liu et al. (1995), the argument is also weighted by the inverse of the scalar ρ

$$(K * \mathbf{u})(\mathbf{x}) = \int_{H_{\mathbf{x}}} \omega\left(\frac{\mathbf{x} - \mathbf{x}'}{\rho}\right) \mathbf{u}\left(\mathbf{x}'\right) \, \mathrm{d}v_{\mathbf{x}'}. \tag{4.119}$$

Since this variant is motivated from the wavelet transformation, ρ is also called dilation. In the least square approach, the difference $\mathbf{a}(\mathbf{x}_I) - \mathbf{u}(\mathbf{x}_I)$ at each node or particle is minimized via the squared error. Using convolution, the identity between both vectors can be described in a weighted integral form

$$\int_{H_{\mathbf{x}}} \omega\left(\frac{\mathbf{x}-\mathbf{x}'}{\rho}\right) \mathbf{a}\left(\mathbf{x}'\right)\, dv_{\mathbf{x}'} = \int_{H_{\mathbf{x}}} \omega\left(\frac{\mathbf{x}-\mathbf{x}'}{\rho}\right) \mathbf{u}\left(\mathbf{x}'\right)\, dv_{\mathbf{x}'}. \tag{4.120}$$

The displacements on the right-hand side can be approximated by $\mathbf{u}\left(\mathbf{x}'\right) = \mathbf{p}\left(\mathbf{x}'\right)\mathbf{d}$ according to Weierstrass theorem. With the dyadic multiplication of both integrands from the left by the vector $\mathbf{p}$ and with the definition of the associated moment tensor $\mathbf{M}$

$$\mathbf{M}(\mathbf{x}) = \int_{H_{\mathbf{x}}} \omega\left(\frac{\mathbf{x}-\mathbf{x}'}{\rho}\right) \mathbf{p}\left(\mathbf{x}'\right) \otimes \mathbf{p}\left(\mathbf{x}'\right)\, dv_{\mathbf{x}'} \tag{4.121}$$

the unknown coefficients $\mathbf{d}$ can be determined from the identity (4.120)

$$\mathbf{d} = \mathbf{M}(\mathbf{x})^{-1} \int_{H_{\mathbf{x}}} \omega\left(\frac{\mathbf{x}-\mathbf{x}'}{\rho}\right) \mathbf{p}\left(\mathbf{x}'\right) \otimes \mathbf{a}\left(\mathbf{x}'\right)\, dv_{\mathbf{x}'}. \tag{4.122}$$

Substituting (4.122) into (4.96), the approximate solution $\mathbf{u}$ in the neighborhood can again be expressed in terms of the coefficients $\mathbf{a}$

$$\mathbf{u}(\mathbf{x}) = \mathbf{p}(\mathbf{x}) \cdot \mathbf{M}^{-1}(\mathbf{x}) \int_{H_{\mathbf{x}}} \omega\left(\frac{\mathbf{x}-\mathbf{x}'}{\rho}\right) \mathbf{p}\left(\mathbf{x}'\right) \mathbf{a}\left(\mathbf{x}'\right)\, dv_{\mathbf{x}'}. \tag{4.123}$$

In the discrete form, (4.123) can be formulated in terms of shape functions

$$\mathbf{u}(\mathbf{x}) = \sum_{J=1}^{N_{\mathbf{x}}} N_J(\mathbf{x})\, \mathbf{a}_J, \quad N_J(\mathbf{x}) = \omega\left(\frac{\mathbf{x}-\mathbf{x}_J}{\rho}\right) \mathbf{p}(\mathbf{x}) \cdot \mathbf{M}^{-1}(\mathbf{x})\, \mathbf{p}(\mathbf{x}_J)\, v_J. \tag{4.124}$$

Remark 1. The difference between the moving least square approximation (4.106) and (4.124) is the multiplication by the volume. Thus, the weighting function ω has a different unit compared to (4.106).

Remark 2. A reproducing kernel (convolution) would exist, if $\mathbf{a}\left(\mathbf{x}'\right) = \mathbf{u}\left(\mathbf{x}'\right)$ in (4.123)

$$\mathbf{u}(\mathbf{x}) = \int_{H_{\mathbf{x}}} K\left(\mathbf{x}, \mathbf{x}'\right) \mathbf{u}\left(\mathbf{x}'\right)\, dv_{\mathbf{x}'}, \quad K\left(\mathbf{x}, \mathbf{x}'\right) = \omega\left(\frac{\mathbf{x}-\mathbf{x}'}{\rho}\right) \mathbf{p}(\mathbf{x}) \cdot \mathbf{M}^{-1}(\mathbf{x})\, \mathbf{p}\left(\mathbf{x}'\right). \tag{4.125}$$

Remark 3. The Wendland function presented in Sect. 4.6.2 is a suitable candidate for a window function due to its properties.

Variant 2. Starting from the idea of deriving shape functions from a linear integral operator, an alternative formulation is presented in Liu et al. (1997). Motivated by the Weierstrass theorem, it is postulated that a function in the neighborhood of a fixed point $\bar{\mathbf{x}}$ can be approximated via a polynomial which argument is shifted and normalized

$$\mathbf{u}(\mathbf{x}, \bar{\mathbf{x}}) = \mathbf{p}\left(\frac{\bar{\mathbf{x}} - \mathbf{x}}{\rho}\right) \mathbf{d}(\bar{\mathbf{x}}) . \tag{4.126}$$

Similar to the least square approach, the error in (4.91) is minimized. In contrast to Sect. 4.6.1, the functional consists of the weighted error integrated over the neighborhood of $\bar{\mathbf{x}}$. The displacement $\mathbf{u}(\mathbf{x}', \bar{\mathbf{x}})$ is approximated using (4.126)

$$J(\mathbf{d}(\bar{\mathbf{x}})) = \int_{H_{\bar{\mathbf{x}}}} \omega\left(\frac{\bar{\mathbf{x}} - \mathbf{x}'}{\rho}\right) \left\| \mathbf{a}(\mathbf{x}') - \mathbf{p}\left(\frac{\bar{\mathbf{x}} - \mathbf{x}'}{\rho}\right) \mathbf{d}(\bar{\mathbf{x}}) \right\|^2 dv_{\mathbf{x}'} \rightarrow \text{min.} \tag{4.127}$$

The computation of the unknown coefficients $\mathbf{d}$ corresponds to the procedure described in Sect. 4.6.1

$$\begin{aligned} \mathbf{d}(\bar{\mathbf{x}}) &= \mathbf{M}^{-1}(\bar{\mathbf{x}}) \int_{H_{\mathbf{x}}} \omega\left(\frac{\bar{\mathbf{x}} - \mathbf{x}'}{\rho}\right) \mathbf{p}\left(\frac{\bar{\mathbf{x}} - \mathbf{x}'}{\rho}\right) \otimes \mathbf{a}(\mathbf{x}') \, dv_{\mathbf{x}'}, \\ \mathbf{M}(\bar{\mathbf{x}}) &= \int_{H_{\mathbf{x}}} \omega\left(\frac{\bar{\mathbf{x}} - \mathbf{x}'}{\rho}\right) \mathbf{p}\left(\frac{\bar{\mathbf{x}} - \mathbf{x}'}{\rho}\right) \otimes \mathbf{p}\left(\frac{\bar{\mathbf{x}} - \mathbf{x}'}{\rho}\right) dv_{\mathbf{x}'}. \end{aligned} \tag{4.128}$$

By substituting (4.128) into (4.126), the solution function can again be described in terms of $\mathbf{a}$

$$\mathbf{u}(\mathbf{x}, \bar{\mathbf{x}}) = \mathbf{p}\left(\frac{\bar{\mathbf{x}} - \mathbf{x}}{\rho}\right) \cdot \mathbf{M}^{-1}(\bar{\mathbf{x}}) \int_{H_{\bar{\mathbf{x}}}} \omega\left(\frac{\bar{\mathbf{x}} - \mathbf{x}'}{\rho}\right) \mathbf{p}\left(\frac{\bar{\mathbf{x}} - \mathbf{x}'}{\rho}\right) \mathbf{a}(\mathbf{x}') \, dv_{\mathbf{x}'}. \tag{4.129}$$

By evaluating at a fixed point in the domain, the above equation corresponds to a weighted least square approach. A moving least square approximation results from the transition $\bar{\mathbf{x}} \rightarrow \mathbf{x}$.

$$\mathbf{u}(\mathbf{x}) = \mathbf{p}(0) \cdot \mathbf{M}^{-1}(\mathbf{x}) \int_{H_{\mathbf{x}}} \omega\left(\frac{\mathbf{x} - \mathbf{x}'}{\rho}\right) \mathbf{p}\left(\frac{\mathbf{x} - \mathbf{x}'}{\rho}\right) \mathbf{a}(\mathbf{x}') \, dv_{\mathbf{x}'}. \tag{4.130}$$

In the discrete form, (4.130) can be represented in terms of shape functions and associated coefficients

$$\mathbf{u}(\mathbf{x}) = \sum_{J=1}^{N_{\mathbf{x}}} N_J(\mathbf{x}) \, \mathbf{a}_J, \quad N_J(\mathbf{x}) = \omega\left(\frac{\mathbf{x} - \mathbf{x}_J}{\rho}\right) \mathbf{p}(0) \cdot \mathbf{M}^{-1}(\mathbf{x}) \, \mathbf{p}\left(\frac{\mathbf{x} - \mathbf{x}_J}{\rho}\right) v_J. \tag{4.131}$$

Remark 1. If $\mathbf{a}(\mathbf{x}') = \mathbf{u}(\mathbf{x}')$, (4.130) corresponds to a mapping with a reproducing kernel (convolution).

Remark 2. Variant 1 results from a shift of the argument in the base function vector by $\mathbf{x}$ and setting $\rho = 1$.

Variant 3. A third variant is given e.g. in Chen et al. (1996). The starting point is to correct the convolution (4.119) so that the function in the integral corresponds to a reproducing kernel

$$\mathbf{u}(\mathbf{x}) = \int_{H_\mathbf{x}} K\left(\mathbf{x}, \mathbf{x}'\right) \mathbf{u}\left(\mathbf{x}'\right) \, \mathrm{d}v_{\mathbf{x}'}, \quad K\left(\mathbf{x}, \mathbf{x}'\right) = C\left(\mathbf{x}, \mathbf{x}'\right) \omega \left(\frac{\mathbf{x} - \mathbf{x}'}{\rho}\right). \tag{4.132}$$

The correction factor C is chosen to depend on a given vector $\widetilde{\mathbf{p}}$ and unknown coefficients

$$C(\mathbf{x}, \mathbf{y}) = \widetilde{\mathbf{p}}\left(\mathbf{x} - \mathbf{x}'\right) \cdot \mathbf{d}(\mathbf{x}). \tag{4.133}$$

The vector $\mathbf{d}$ is determined such that the kernel K has a reproducing property. The vector $\mathbf{u}$ within the integral of (4.132) can also be expressed by a Taylor series around $\mathbf{x}$

$$\mathbf{u}\left(\mathbf{x}'\right) = \mathbf{u}(\mathbf{x}) + \frac{\partial \mathbf{u}(\mathbf{x})}{\partial \mathbf{x}} \left(\mathbf{x} - \mathbf{x}'\right) + \mathcal{O}\left(\left(\mathbf{x} - \mathbf{x}'\right)^2\right). \tag{4.134}$$

By substituting (4.134) into (4.132), conditions for a reproducible kernel can be established

$$\begin{aligned} \mathbf{u}(\mathbf{x}) = & \int_{H_\mathbf{x}} \widetilde{\mathbf{p}}\left(\mathbf{x} - \mathbf{x}'\right) \cdot \mathbf{d}(\mathbf{x}) \, \omega \left(\frac{\mathbf{x} - \mathbf{x}'}{\rho}\right) \\ & \left(\mathbf{u}(\mathbf{x}) + \frac{\partial \mathbf{u}(\mathbf{x})}{\partial \mathbf{x}} \left(\mathbf{x} - \mathbf{x}'\right) + \mathcal{O}\left(\left(\mathbf{x} - \mathbf{x}'\right)^2\right)\right) \mathrm{d}v_{\mathbf{x}'}. \end{aligned} \tag{4.135}$$

The equal sign in (4.135) is satisfied if and only if the first term equals to one

$$\int_{H_\mathbf{x}} \widetilde{\mathbf{p}}\left(\mathbf{x} - \mathbf{x}'\right) \cdot \mathbf{d}(\mathbf{x}) \, \omega \left(\frac{\mathbf{x} - \mathbf{x}'}{\rho}\right) \mathrm{d}v_{\mathbf{x}'} = 1, \tag{4.136}$$

and the other terms to zero

$$\int_{H_\mathbf{x}} \widetilde{\mathbf{p}}\left(\mathbf{x} - \mathbf{x}'\right) \cdot \mathbf{d}(\mathbf{x}) \, \omega \left(\frac{\mathbf{x} - \mathbf{x}'}{\rho}\right) \left(\mathbf{x} - \mathbf{x}'\right)^k \mathrm{d}v_{\mathbf{x}'} = 0, \quad \text{for each} \quad k = 1, \ldots, n. \tag{4.137}$$

The quantity n corresponds to the number of terms in the Taylor series that contain a derivative larger or equal to one. Both conditions can also be combined. In (4.136) the expression $\left(\mathbf{x} - \mathbf{x}'\right)^0$ can be added. The last factor in the integral on the right-hand side can be expressed by the base function vector $\mathbf{p}$. The same applies to the right-hand side. In this case, the function is evaluated at the point $\mathbf{x} = \mathbf{x}'$

$$\int_{H_{\mathbf{x}}} \widetilde{\mathbf{p}}\left(\mathbf{x}-\mathbf{x}'\right)\cdot \mathbf{d}\left(\mathbf{x}\right)\omega\left(\frac{\mathbf{x}-\mathbf{x}'}{\rho}\right)\mathbf{p}\left(\mathbf{x}-\mathbf{x}'\right)\, dv_{\mathbf{x}'} = \mathbf{p}\left(\mathbf{x}-\mathbf{x}'\right)\Big|_{\mathbf{x}=\mathbf{x}'} = \mathbf{p}\left(0\right). \tag{4.138}$$

Here it is important to note that the number of terms in the Taylor series (4.134) must equal the number of terms in $\mathbf{p}$. From the evaluation of (4.138), the unknown coefficients can be determined

$$\mathbf{d}\left(\mathbf{x}\right) = \mathbf{M}^{-1}\left(\mathbf{x}\right)\mathbf{p}\left(0\right), \quad \mathbf{M}\left(\mathbf{x}\right) = \int_{H_{\mathbf{x}}} \omega\left(\frac{\mathbf{x}-\mathbf{x}'}{\rho}\right)\widetilde{\mathbf{p}}\left(\mathbf{x}-\mathbf{x}'\right)\otimes\mathbf{p}\left(\mathbf{x}-\mathbf{x}'\right)\, dv_{\mathbf{x}'}. \tag{4.139}$$

By substituting (4.139) into (4.132) and taking advantage of (4.133), a similar formulation as in variant two (4.130) results

$$\mathbf{u}\left(\mathbf{x}\right) = \mathbf{p}\left(0\right)\cdot\mathbf{M}^{-T}\left(\mathbf{x}\right)\int_{H_{\mathbf{x}}} \omega\left(\frac{\mathbf{x}-\mathbf{x}'}{\rho}\right)\widetilde{\mathbf{p}}\left(\mathbf{x}-\mathbf{x}'\right)\mathbf{u}\left(\mathbf{x}'\right)\, dv_{\mathbf{x}'}. \tag{4.140}$$

Also the correction factor can now be specified in detail

$$C\left(\mathbf{x},\mathbf{x}'\right) = \mathbf{p}\left(0\right)\cdot\mathbf{M}^{-T}\left(\mathbf{x}\right)\widetilde{\mathbf{p}}\left(\mathbf{x}-\mathbf{x}'\right) \tag{4.141}$$

If the base function vector is chosen in (4.133), i.e. $\widetilde{\mathbf{p}} = \mathbf{p}$, then both formulations are equivalent. All three variants have in common that the shape functions fulfill the reproducing conditions up to the desired order.

Remark. Since the third variant is based on a correction, shape function derivatives can be directly computed without any differentiation (Li and Liu 1998). In this context, the term implicit gradient is used. A corrected kernel can be designed to map the k-th-derivative of the function $\mathbf{u}$

$$\frac{\partial^k \mathbf{u}\left(\mathbf{x}\right)}{\partial^k \mathbf{x}} = \int_{H_{\mathbf{x}}} C\left(\mathbf{x},\mathbf{x}'\right)\omega\left(\frac{\mathbf{x}-\mathbf{x}'}{\rho}\right)\mathbf{u}\left(\mathbf{x}'\right)\, dv_{\mathbf{x}'}. \tag{4.142}$$

Exploiting the Taylor series (4.134) together with the approach for the correction (4.133) an equation can be constructed to determine the unknown coefficients $\mathbf{d}$

$$\begin{aligned}\frac{\partial^k \mathbf{u}\left(\mathbf{x}\right)}{\partial^k \mathbf{x}} = &\int_{H_{\mathbf{x}}} \widetilde{\mathbf{p}}\left(\mathbf{x}-\mathbf{x}'\right)\cdot\mathbf{d}\left(\mathbf{x}\right)\omega\left(\frac{\mathbf{x}-\mathbf{x}'}{\rho}\right)\\ &\left(\mathbf{u}\left(\mathbf{x}\right) + \frac{\partial \mathbf{u}\left(\mathbf{x}\right)}{\partial \mathbf{x}}\left(\mathbf{x}-\mathbf{x}'\right) + \mathcal{O}\left(\left(\mathbf{x}-\mathbf{x}'\right)^2\right)\right)\, dv_{\mathbf{x}'}.\end{aligned} \tag{4.143}$$

The equal sign only holds, if the relation

$$\int_{H_{\mathbf{x}}} \widetilde{\mathbf{p}}\left(\mathbf{x}-\mathbf{x}'\right)\cdot\mathbf{d}\left(\mathbf{x}\right)\omega\left(\frac{\mathbf{x}-\mathbf{x}'}{\rho}\right)\mathbf{p}\left(\mathbf{x}-\mathbf{x}'\right)\, dv_{\mathbf{x}'} = \frac{1}{k!}\frac{\partial^k \mathbf{p}\left(\mathbf{x}-\mathbf{x}'\right)}{\partial^k \mathbf{x}}\Bigg|_{\mathbf{x}=\mathbf{x}'} \tag{4.144}$$

is satisfied. From this equation the unknown coefficients can be calculated. Substituting **d** into (4.142) yields the approximation of the k-th derivative

$$\frac{\partial^k \mathbf{u}(\mathbf{x})}{\partial^k \mathbf{x}} = \frac{1}{k!} \frac{\partial^k \mathbf{p}(\mathbf{x}-\mathbf{x}')}{\partial^k \mathbf{x}}\bigg|_{\mathbf{x}=\mathbf{x}'} \cdot \mathbf{M}^{-T}(\mathbf{x}) \int_{H_\mathbf{x}} \omega\left(\frac{\mathbf{x}-\mathbf{x}'}{\rho}\right) \widetilde{\mathbf{p}}(\mathbf{x}-\mathbf{x}')\,\mathbf{u}(\mathbf{x}')\;\mathrm{d}v_{\mathbf{x}'}. \tag{4.145}$$

The only difference between (4.145) and (4.140) is the first factor that can be easily calculated beforehand. As noted in Hillman et al. (2020), the implicit gradient approach is equivalent to the peridynamic derivatives from Madenci et al. (2016).

References

N. Aronszajn, Theory of reproducing kernels. Trans. Am. Math. Soc. **68**(3), 337–404 (1950)

M. Arroyo, M. Ortiz, Local maximum-entropy approximation schemes: a seamless bridge between finite elements and meshfree methods. Int. J. Num. Methods Eng. **65**, 2167–2202 (2006)

K.J. Bathe, *Finite Element Procedures* (Klaus-Jurgen Bathe, 2006)

G.P. Bazeley, Y.K. Cheung, B.M. Irons, O.C. Zienkiewicz, Triangular elements in bending-conforming and nonconforming solutions, in *Proceedings of the Conference on Matrix Methods in Structural Mechanics, Air Forces Institute of Technology, Wright Patterson AF Base, Ohio* (1965)

T. Belytschko, Y. Krongauz, J. Dolbow, C. Gerlach, On the completeness of meshfree particle methods. Int. J. Num. Methods Eng. **43**(5), 785–819 (1998)

T. Belytschko, Y. Guo, W.K. Liu, S.P. Xiao, A unified stability analysis of meshless particle methods. Int. J. Numer. Methods Eng. **48**(9), 1359–1400 (2000a)

T. Belytschko, W.K. Liu, B. Moran. *Nonlinear Finite Elements for Continua and Structures* (Wiley, Chichester, 2000b)

J. Bonet, S.D. Kulasegaram, Correction and stabilization of Smooth Particle Hydrodynamics methods with applications in metal forming simulations. Int. J. Num. Methods Eng. **47**(6), 1189–1214 (2000)

J.U. Brackbill, D.B. Kothe, C. Zemach, A continuum method for modeling surface tension. J. Comput. Phys. **100**, 335–354 (1992)

D. Braess. *Finite Elemets: Theory, Fast Solvers, and Applications in Solid Mechanics*, 3rd edn. (Cambridge University Press, 2007)

A.N. Brooks, T.J.R. Hughes, Streamline upwind/Petrov-Galerkin formulations for convection dominated flows with particular emphasis on the incompressible Navier-Stokes equations. Comput. Methods Appl. Mech. Eng. **32**(1–3), 199–259 (1982)

J.S. Chen, C. Pan, C.T. Wu, W.K. Liu, Reproducing kernel particle methods for large deformation analysis of non-linear structures. Comput. Methods Appl. Mech. Eng. **139**(1–4), 195–227 (1996)

J.S. Chen, C.T. Wu, S. Yoon, Y. You, A stabilized conforming nodal integration for Galerkin meshfree methods. Int. J. Num. Methods Eng. **50**(2), 435–466 (2001)

J.S. Chen, S. Yoon, C.T. Wu, Non-linear version of stabilized conforming nodal integration for Galerkin mesh-free methods. Int. J. Numer. Methods Eng. **53**(12), 2587–2615 (2002)

J.S. Chen, M. Hillman, M. Rüter, An arbitrary order variationally consistent integration for Galerkin meshfree methods. Int. J. Numer. Methods Eng. **95**(5), 387–418 (2013)

L. Chen, J.H. Lee, C.F. Chen, On the modeling of surface tension and its applications by the generalized interpolation material point method. Comput. Model. Eng. Sci. **86**(3), 199 (2012)

G. Dhatt, G. Touzot, *The Finite Element Method Displayed* (Wiley, Chicester, 1984)

I. Ergatoudis, B.M. Irons, O.C. Zienkiewicz, Curved, isoparametric, "quadrilateral" elements for finite element analysis. Int. J. Solids Struct. **4**(1), 31–42 (1968)

W. Fleming. *Functions of Several Variables* (Springer Science & Business Media, 2012)
M. Foca. *On a Local Maximum Entropy Interpolation Approach for Simulation of Coupled Thermo-mechanical Problems. Application to the Rotary Frictional Welding process*. Ph.D. thesis, Ecole Centrale de Nantes (ECN) (2015)
P.M. Gresho, R.L. Sani, *Incompressible Flow and the Finite Element Method. Volume 1: Advection-diffusion and Isothermal Laminar Flow* (Wiley, New York, NY (USA), 1998)
M. Hillman, M. Pasetto, G. Zhou, Generalized reproducing kernel Peridynamics: unification of local and non-local meshfree methods, non-local derivative operations, and an arbitrary-order state-based peridynamic formulation. Comput. Part. Mech. **7**(2), 435–469 (2020)
C.W. Hirt, B.D. Nichols, Volume of fluid (VOF) method for the dynamics of free boundaries. J. Comput. Phys. **39**(1), 201–225 (1981)
G.A. Holzapfel, *Nonlinear Solid Mechanics. A Continuum Approach for Engineering* (Wiley, Chichester, 2000)
T.R.J. Hughes, *The Finite Element Method* (Prentice Hall, Englewood Cliffs, NJ, 1987)
B.M. Irons, O.C. Zienkiewicz, The isoparametric finite element system: a new concept in finite element analysis. *Royal Aeronautical Society London* 3–36 (1968)
E.T. Jaynes, Information theory and statistical mechanics. Phys. Rev. **106**(4), 620–630 (1957)
Y. Krongauz, T. Belytschko, Consistent pseudo-derivatives in meshless methods. Comput. Methods Appl. Mech. Eng. **146**(3–4), 371–386 (1997)
S. Kumar, K. Danas, D.M. Kochmann, Enhanced local maximum-entropy approximation for stable meshfree simulations. Comput. Methods Appl. Mech. Eng. **344**, 858–886 (2019)
P. Lancaster, K. Salkauskas, *Curve and Surface Fitting. An Introduction* (Academic Press, London, 1986)
S. Li, W.K. Liu, Synchronized reproducing kernel interpolant via multiple wavelet expansion. Comput. Mech. **21**(1), 28–47 (1998)
S. Li, W.K. Liu, *Meshfree Particle Methods* (Springer, Berlin, Heidelberg, New York, 2007)
I.V. Lindell, *Methods for Electromagnetic Field Analysis* (IEEE Press, 1992)
W.K. Liu, S. Jun, Y.F. Zhang, Reproducing kernel particle methods. Int. J. Numer. Methods Fluids **20**(8–9), 1081–1106 (1995)
W.K. Liu, S. Li, T. Belytschko, Moving least-square reproducing kernel methods (I): methodology and convergence. Comput. Methods Appl. Mech. Eng. **143**(1), 113–154 (1997)
E. Madenci, A. Barut, M. Futch, Peridynamic differential operator and its applications. Compu. Methods Appl. Mech. Eng. **304**, 408–451 (2016)
E. Madenci, M. Dorduncu, A. Barut, N. Phan, Weak form of Peridynamics for nonlocal essential and natural boundary conditions. Comput. Methods Appl. Mech. Eng. **337**, 598–631 (2018)
J. Marsden, T.J.R. Hughes, *Mathematical Foundations of Elasticity* (Prentice-Hall Inc, Englewood Cliffs, 1983)
C. Miehe, *Zur numerischen Behandlung thermomechanischer Prozesse*. Ph.D. thesis, Universität Hannover, Germany (1988)
E. Onate, S. Idelsohn, O.C. Zienkiewicz, R.L. Taylor, A finite point method in computational mechanics. Applications to convective transport and fluid flow. *International Journal for Numerical Methods in Engineering* **39**, 3839–3866 (1996)
S. Osher, R.P. Fedkiw, Level set methods: an overview and some recent results. J. Comput. Phys. **169**(2), 463–502 (2001)
S. Osher, J.A. Sethian, Fronts propagating with curvature-dependent speed: algorithms based on Hamilton-Jacobi formulations. J. Comput. Phys. **79**(1), 12–49 (1988)
D.J. Price, Smoothed particle hydrodynamics and magnetohydrodynamics. J. Comput. Phys. **231**(3), 759–794 (2012)
M.A. Puso, J.S. Chen, E. Zywicz, W. Elmer, Meshfree and finite element nodal integration methods. Int. J. Numer. Methods Eng. **74**(3), 416–446 (2008)
P.W. Randles, L.D. Libersky, Smoothed particle hydrodynamics: some recent improvements and applications. Comput. Methods Appl. Mech. Eng. **139**(1–4), 375–408 (1996)

R. Seydel, *Practical Bifurcation and Stability Analysis*, vol. 5 (Springer Science & Business Media, 2009)

J.C. Simo, N. Tarnow, The discrete energy-momentum method. Conserving algorithms for nonlinear elastodynamics. *Zeitschrift für angewandte Mathematik und Physik ZAMP* **43**(5), 757–792 (1992)

G. Strang, Variational crimes in the Finite Element Method, in *The Mathematical Foundations of the Finite Element Method with Applications to Partial Differential Equations* (Elsevier, 1972), pp. 689–710

G. Strang, G.J. Fix, *An Analysis of the Finite Element Methhod* (Prentice-Hall, Englewood Cliffs, NJ, 1973)

J.W. Swegle, D.L. Hicks, S.W. Attaway, Smoothed particle hydrodynamics stability analysis. J. Comput. Phys. **116**(1), 123–134 (1995)

R.L. Taylor, J.C. Simo, O.C. Zienkiewicz, A.C.H. Chan, The patch test—a condition for assessing FEM convergence. Int. J. Numer. Methods Eng. **22**(1), 39–62 (1986)

M.R. Tupek, R. Radovitzky, An extended constitutive correspondence formulation of Peridynamics based on nonlinear bond-strain measures. J. Mech. Phys. Solids **65**, 82–92 (2014)

M.M. Vainberg, *Variational Methods for the Study of Nonlinear Operators* (Holden Day, 1964)

C. Weißenfels, Direct nodal imposition of surface loads using the divergence theorem. Finite Elem. Anal. Des. **165**, 31–40 (2019)

H. Wendland, Piecewise polynomial, positive definite and compactly supported radial functions of minimal degree. Adv. Comput. Math. **4**(1), 389–396 (1995)

W.L. Wood, *Practical Time-Stepping Schemes*, vol. 6 (Clarendon Press, Oxford, 1990)

P. Wriggers. *Nonlinear Finite Element Methods* (Springer Science & Business Media, 2008)

O.C. Zienkiewicz, R.L. Taylor, *The Finite Element Method*, vol. 1, 4th edn. (McGraw Hill, London, 1989)

Chapter 5
Meshfree Galerkin Methods

The Finite Element Method based on a Lagrangian description of the differential equation is mostly used to simulate the behavior of solids under loadings. With this scheme, a good approximation of the solution can be achieved, provided that the elements do not distort too much. However, this restriction limits the range of applications. To increase the flexibility of Galerkin methods, approaches are pursued which allow the determination of the test and trial function on an almost arbitrary distribution of nodes in the corresponding neighborhood. A variety of meshfree methods based on the weak form were developed over the years. In the context of this chapter, two methods are presented in more detail. The Reproducing Kernel Particle Method (RKPM) exemplifies one type of meshfree Galerkin scheme. Due to numerous developments in the last years, which are especially based on this method, the RPKM is investigated with respect to the fulfillment of the criteria for spatial discretization schemes. The Optimal Transportation Meshfree (OTM) method represents a relatively new procedure and is therefore examined in more detail. In addition to these, other discretization schemes based on the weak form exist. An overview can be found, for example, in Li and Liu (2007).

5.1 Reproducing Kernel Particle Method

A first attempt to resolve the rigid subdivision of the domain into elements led to the Diffuse Element Method (Nayroles et al. 1992). The approach to approximate the trial and test function is analogous to the Finite Element Method. However, the domain is not subdivided into elements and the number of nodes N_p^{sup} in the support domain of an integration point can now be arbitrary (Fig. 5.1). The shape functions are based on the moving least square concept from Sect. 4.6.1. However, the weak

C. Weißenfels, *Simulation of Additive Manufacturing using Meshfree Methods*,
Lecture Notes in Applied and Computational Mechanics 97,
https://doi.org/10.1007/978-3-030-87337-0_5

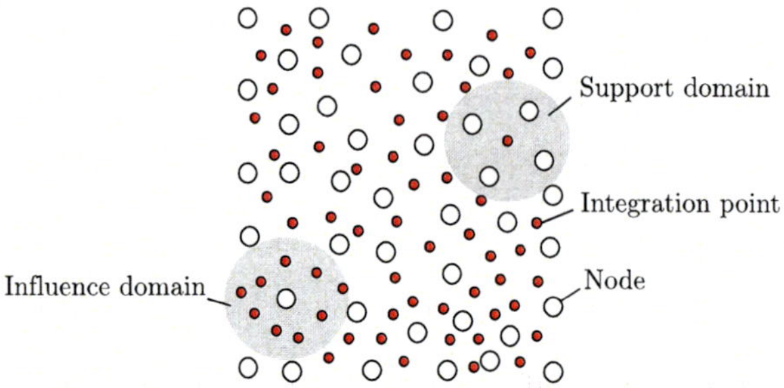

Fig. 5.1 Subdivision of the domain into nodes and integration points in case of a two-point meshfree Galerkin method

form is not evaluated with respect to local orthonormal reference elements, but at global coordinates.

The inconsistency in the derivative of the shape function was later corrected in Belytschko et al. (1994). In addition, the name of the scheme was changed to Element Free Galerkin Method. Often, instead of the classical moving least square functions, the corresponding equivalents are applied, which derive from integral operators. When using the first or third variant from Sect. 4.6.4, the name Reproducing Kernel Particle Method (Liu et al. 1995) has become established. The application of the second variant from Sect. 4.6.4 leads to the Moving Least-Square Reproducing Kernel Method (Liu et al. 1997). All the shape functions that are used in these schemes satisfy the reproducing conditions from Sect. 4.5.2.

Two-point method. In the first publications, the Gauss quadrature rule is applied to evaluate the volume integral. In Belytschko et al. (1994) the domain is subdivided into individual cells. Each cell consists of a certain number of integration points n_{cell}. This value is determined by the number of nodes m in the cell, like e.g. for the 2-dimensional case

$$n_{cell} = n_Q \times n_Q, \qquad n_Q = \sqrt{m} + 2. \tag{5.1}$$

The number of cells m_c is related to the total number of nodes n in the domain $m_c = \sqrt{n}$. The applications focus on simulations to predict the deformation behavior of solids. Therefore, the equations are usually formulated with respect to the coordinates of the initial configuration. The trial and test function at the integration point can be approximated from the nodal values in the support domain

$$\mathbf{u}(\mathbf{X}_p) = \sum_{I=1}^{N_p^{sup}} N_I(\mathbf{X}_p)\mathbf{a}_I, \qquad \eta(\mathbf{X}_p) = \sum_{I=1}^{N_p^{sup}} N_I(\mathbf{X}_p)\mathbf{b}_I. \tag{5.2}$$

The actual force at a node is calculated analogously to (4.42) with the evaluation of the individual quantities with respect to the initial configuration

$$\mathbf{f}_{I\,n+1} = \sum_{p=1}^{N_I^{inf}} \boldsymbol{\sigma}_{n+1}(\mathbf{X}_p)\,\frac{\partial N_I(\mathbf{X}_p)}{\partial \mathbf{x}_{p\,n+1}}\,v_{n+1}(\mathbf{X}_p).$$

One-point method. Using background cells to locate integration points does not correspond to a true meshfree method. In order to increase the flexibility and to reduce the computational efforts, an evaluation of the integral at the nodes has become more popular. In this case, (5.2) changes to

$$\mathbf{u}(\mathbf{X}_I) = \sum_{J=1}^{N_I} N_I(\mathbf{X}_J)\mathbf{a}_J, \qquad \eta(\mathbf{X}_I) = \sum_{J=1}^{N_I} N_I(\mathbf{X}_J)\mathbf{b}_J. \tag{5.3}$$

The neighborhoods for approximating the trial as well as the test function, and thus for computing the force at a node, are identical

$$\mathbf{f}^s_{I\,n+1} = \sum_{J=1}^{N_I} \boldsymbol{\sigma}_{n+1}(\mathbf{X}_J)\,\frac{\partial N_I(\mathbf{X}_J)}{\partial \mathbf{x}_{J\,n+1}}\,v_{n+1}(\mathbf{X}_J). \tag{5.4}$$

Here it is important to note that the abbreviation N_I above the sum sign denotes the number of nodes in the neighborhood of node I.

5.1.1 Search Algorithm

For most methods, the neighborhoods are determined from a search algorithm. In Liu et al. (1997) or in Li and Liu (2007), criteria are listed, which must be fulfilled for an admissible node or particle distribution, see also Sect. 4.5.8. The equations are mostly formulated with respect to the initial configuration. The additional condition that the force at a node must be identical before and after the search is thus omitted in this case.

5.1.2 Rank Instability

The moving least square functions are rational. For a sufficiently accurate integration, a correspondingly large number of evaluation points is required. In De and Bathe (2001), for example, it is shown in the context of the Finite Sphere Method that 144 integration points are needed in the neighborhood to achieve a convergent response in the two-dimensional case.

The evaluation of the integral at the nodes limits the total number. In addition, the derivative of the shape function at the node I of the neighborhood H_I of I is equal to zero. As a consequence, artificial zero-energy modes result (Beissel and Belytschko 1996). A detailed analysis for the 1-dimensional case can be found in Puso et al. (2008) or Hillman and Chen (2016). Over the years, different approaches to stabilize the negative effects due to under-integration emerged. Two different concepts are briefly presented.

Strain smoothing stabilization. Chen et al. (2001) proposes an alternative derivative rules based on averaging. For the case of large deformations (Chen et al. 2002), the derivatives are determined from the averaged gradient in the neighborhood. By using Gauss's integral theorem, this quantity can also be obtained by evaluating a surface integral

$$\overline{\text{Grad}\,\mathbf{u}} = \frac{1}{vol(H_{\mathbf{X}})}\int_{H_{\mathbf{X}}} \text{Grad}\,\mathbf{u}\;\text{d}V = \frac{1}{vol(H_{\mathbf{X}})}\int_{\partial H_{\mathbf{X}}} \mathbf{u}\otimes\mathbf{N}\;\text{d}A. \tag{5.5}$$

The bar above a quantity denotes its averaged value. From the discretization of (5.5) using (4.1) an averaged derivative of the shape function can be constructed

$$\overline{\text{Grad}\,\mathbf{u}\,(\mathbf{X}_I)} = \sum_{J=1}^{N_I} \mathbf{a}_J \otimes \frac{\partial \bar{N}_J\,(\mathbf{X}_I)}{\partial \mathbf{X}_I} = \frac{1}{vol(H_{\mathbf{X}_I})}\int_{\Gamma_I} \sum_{J=1}^{N_I} \mathbf{a}_J \otimes N_J\,(\mathbf{X})\,\mathbf{N}\,(\mathbf{X})\;\text{d}A. \tag{5.6}$$

By neglected the components the individual derivative can be calculated from the evaluation of a boundary integral

$$\frac{\partial \bar{N}_J\,(\mathbf{X}_I)}{\partial \mathbf{X}_I} = \frac{1}{vol(H_{\mathbf{X}_I})}\int_{\Gamma_I} N_J\,(\mathbf{X})\,\mathbf{N}\,(\mathbf{X})\;\text{d}A, \quad vol(H_{\mathbf{X}_I}) = \int_{H_{\mathbf{X}_I}} \text{d}V. \tag{5.7}$$

To determine the normalized normal vector $\mathbf{N}$ and to evaluate the integral, it is necessary to know the surface Γ_I belonging to node I. Therefore, in Chen et al. (2001) the domain is uniquely decomposed into individual Voronoi cells. Thus, this approach, called Stabilized Conforming Nodal Integration (SCNI), no longer corresponds to a true meshfree method. To preserve the smoothing or averaging character for stabilization, and to avoid a unique decomposition of the domain into individual cells, alternatively, the Stabilized Non-Conforming Nodal Integration (SNNI) approach (Chen et al. 2007) can be applied. Here, each node is assigned its own cell, which can be spherical or cuboidal. The decomposition is no longer unique and the individual cells may overlap.

In Puso et al. (2008) it is shown that both approaches are not sufficient to stabilize the negative effects due to under-integration, see also Hillman and Chen (2016). Especially when the surface-to-volume ratio is small or when the domain is resolved very finely, artificial zero-energy modes appear even with this scheme. Therefore, in Chen et al. (2007), see also Puso et al. (2008), it is recommended to evaluate the weak form additionally at locations inside the cells. Therefore, the cells can be further subdivided into triangles, for instance.

Gradient based stabilization. To avoid rank instability, an additional term G^{stab} can be added to the weak form as suggested in Beissel and Belytschko (1996). In terms of a penalty formulation, the strong form of the differential equation is enforced to be satisfied

$$G^{stab}(\mathbf{u}, \eta) = c \int_{B_t} \operatorname{div} \mathbf{q} \cdot \left(\operatorname{div} \boldsymbol{\sigma} + \rho \bar{\mathbf{b}} \right) \, dv. \tag{5.8}$$

The penalty parameter c must be determined appropriately before the calculation. For the additional test function $\mathbf{q}$, an analogous approach as for η can be made. In G^{stab}, the second derivative of the shape function occurs. Since this value is nonzero at the central node of the neighborhood, a stabilizing effect can be obtained.

A similar but simplified approach, which is more suitable for the non-linear case, can be found in Puso et al. (2008). Instead of the strong form, an additional term based on the deformation gradient and the second derivative of the shape function is added.

Another stabilization is presented in Hillman and Chen (2016). Therein, the test and trial functions are evaluated not only at one point but in the whole neighborhood. The discrete virtual internal work G^i from (4.12) is calculated alternatively

$$G^i(\eta, \mathbf{u}) \approx \sum_{J=1}^{n} \int_{H_{\mathbf{X}_J}} \operatorname{grad} \eta (\mathbf{X}_J, \mathbf{X}) \cdot \boldsymbol{\sigma} (\mathbf{x}_J, \mathbf{x}) \, dV. \tag{5.9}$$

The test function is determined from a Taylor series around the location $\mathbf{X}_J$, which is terminated after the linear term

$$\eta (\mathbf{X}_J, \mathbf{X}) = \sum_{I=1}^{N_J} N_I(\mathbf{X}_J) \mathbf{b}_I + \sum_{I=1}^{N_I} \frac{\partial N_I(\mathbf{X}_J)}{\partial \mathbf{X}_J} \cdot (\mathbf{X} - \mathbf{X}_J) \, \mathbf{u}_J. \tag{5.10}$$

The trial function is discretized in an analogous way. The advantage of this approach is not only the reduction of computational efforts compared to the previous corrections. Moreover, it does not require any additional numerical parameter. When using reproducing kernel functions, the derivative of the shape function in (5.10) can be determined directly in terms of implicit gradients (4.145).

5.1.3 Integration Constraint

In Dolbow and Belytschko (1999) it is shown that for a two-point subdivision of the domain the integration error depends not only on the number but also on the location of the integration points. A proper adjustment of the background cells to the support domains leads to an improved solution. The integration constraint determines the optimal location. Two approaches, which are developed in the context of a one-point method, are presented to satisfy this requirement.

Conforming nodal integration. The unique decomposition of the domain into Voronoi cells in the framework of Stabilized Conforming Nodal Integration together with the shape function derivatives based on averaging ensures that the integration constraint is always satisfied (Chen et al. 2001). Integrating (5.7) leads to (4.77)

$$\int_{H_{\mathbf{X}_J}} \frac{\partial \bar{N}_J(\mathbf{X})}{\partial \mathbf{X}} \, \mathrm{d}V = \sum_{I=1}^{N_J} \frac{\partial \bar{N}_J(\mathbf{X}_I)}{\partial \mathbf{X}_I} V_I = \sum_{I=1}^{N_J} \frac{1}{vol(H_{\mathbf{X}_I})} \left(\int_{\Gamma_I} N_J(\mathbf{X}) \, \mathbf{N} \, \mathrm{d}A \right) V_I$$
$$= \sum_{I=1}^{N_J} \int_{\Gamma_I} N_J(\mathbf{X}) \, \mathbf{N} \, \mathrm{d}A = \int_{\partial H_{\mathbf{X}_J}} N_J(\mathbf{X}) \, \mathbf{N} \, \mathrm{d}A. \tag{5.11}$$

The above relation is only valid because the domain is uniquely subdivided into individual sections. In this case, the volume of a section corresponds exactly to the volume of a node, i.e. $vol(H_{\mathbf{X}_I}) = V_I$. Secondly, the nodes in the close environment share a common interface. However, the normal vectors point in the opposite direction. During summation, the surface integrals cancel at interior nodes. A detailed explanation can be found in Chen et al. (2001).

Modified test function. The SCNI approach does not belong to a true meshfree method due to the subdivision of the domain into non overlapping Voronoi cells. Another possibility consists in the correction of the test shape function. This approach is borrowed from SPH (Bonet and Kulasegaram 2000) and can be used for methods that are not based on a unique subdivision of the domain. The derivative of the test function is thereby enriched by an additional contribution

$$\frac{\partial \tilde{N}_I(\mathbf{X})}{\partial \mathbf{X}} = \frac{\partial N_I(\mathbf{X})}{\partial \mathbf{X}} + \sum_{i=1}^{d} \frac{\partial \psi_I(\mathbf{X})}{\partial X_i} \beta_i. \tag{5.12}$$

The abbreviation d denotes the dimension. A suitable choice must be made for the derivative of the function ψ. Different variants can be found in Bonet and Kulasegaram (2000), Puso et al. (2008) or Chen et al. (2013). The unknown coefficients $\boldsymbol{\beta}$ are determined such that the integration constraint is satisfied for each node. An application of this correction in conjunction with Peridynamics can be found in Bode et al. (2021).

5.2 Optimal Transportation Meshfree Method

The Optimal Transportation Meshfree (OTM) method (Li et al. 2010) is understood as a combination of the optimum theory with meshfree shape functions and a material point sampling. An introduction to optimum theory can be found in Villani (2013). In Weißenfels and Wriggers (2018) it is shown that the discrete equations can also be derived from the weak form. Analogous to the approaches from the previous section, the test and trial function at the integration point are computed from the coefficients at the nodes in the support domain. The OTM method according to Li et al. (2010) is based on an updated Lagrangian formulation. Thus, the current values are calculated from the shape functions of the previous load or time step

$$\mathbf{u}_{p\,n+1} = \sum_{I=1}^{N_{p\,n}^{sup}} N_I(\mathbf{x}_{p\,n})\mathbf{a}_{I\,n+1}, \qquad \boldsymbol{\eta}_{p\,n+1} = \sum_{I=1}^{N_{p\,n}^{sup}} N_I(\mathbf{x}_{p\,n})\mathbf{b}_{I\,n+1}. \tag{5.13}$$

In the OTM method, integration points are also called material points. To ensure consistency in the formulation, only the term integration point is used. The actual force at a node is calculated analogously to the equations in Sect. 4.4.3

$$\mathbf{f}_{I\,n+1} = \sum_{p=1}^{N_{I\,n}^{inf}} \boldsymbol{\sigma}_{p\,n+1} \frac{\partial N_I(\mathbf{x}_{p\,n})}{\partial \mathbf{x}_{p\,n+1}} v_{p\,n+1}. \tag{5.14}$$

The number of integration points $N_{I\,n}^{inf}$ in the influence domain of node I results directly from the support domains. The derivative of the shape function with respect to the current coordinates, and the current volume

$$\frac{\partial N_I(\mathbf{x}_{p\,n})}{\partial \mathbf{x}_{p\,n+1}} = \frac{\partial N_I(\mathbf{x}_{p\,n})}{\partial \mathbf{x}_{p\,n}} \Delta\mathbf{F}_{p\,n+1}^{-1}, \qquad v_{p\,n+1} = \det \Delta\mathbf{F}_{p\,n+1} v_{p\,n}$$

are determined via the incremental deformation gradient

$$\Delta\mathbf{F}_{p\,n+1} = \mathbf{1} + \sum_{I=1}^{N_{p\,n}^{sup}} (\mathbf{a}_{I\,n+1} - \mathbf{a}_{I\,n}) \otimes \frac{\partial N_I(\mathbf{x}_{p\,n})}{\partial \mathbf{x}_{p\,n}}, \quad \mathbf{F}_{p\,n+1} = \mathbf{F}_{p\,n}\Delta\mathbf{F}_{p\,n+1}.$$

The current nodal coordinates result directly from the computed coefficients

$$\mathbf{x}_{I\,n+1} = \mathbf{X}_I + \mathbf{a}_{I\,n+1}. \tag{5.15}$$

Since the constitutive relations are nonlinear within the examples of the following sections, the solution scheme presented in Sect. 4.3 is used to the compute the unknown coefficients. The main difference of the OTM method compared to the approaches of the previous section is the update of the integration points. These points move relative to the nodal coordinates, and the new location can be estimated using the shape functions

$$\mathbf{x}_{p\,n+1} = \sum_{I=1}^{N_{p\,n}^{sup}} N_I\left(\mathbf{x}_{p\,n}\right) \mathbf{x}_{I\,n+1}. \tag{5.16}$$

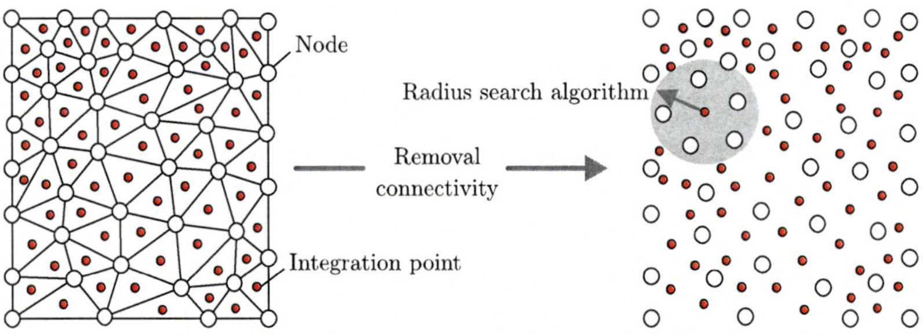

Fig. 5.2 Initial subdivision of the domain into triangles in 2D. Removal of mesh results in location of integration points and nodes

After these updates, the support domains and the shape functions are redefined. Before starting the calculation, the initial coordinates of the integration points and the nodes must be determined. Often, the domain is divided into triangles (2D) or tetrahedra (3D). Afterwards, the element connection list is jettisoned (see Fig. 5.2).

The OTM method in this form does not fulfill all requirements of Sect. 4.5. In addition to presenting possible algorithms for determining the support domains, the following sections provide an insight into the challenges of this method. In the end, an alternative algorithm is presented that satisfies the integration constraint together with the reproducing conditions for the trial functions. All investigations are limited to the quasi-static case to avoid influences from the inertia term.

Remark Since the OTM method can be derived from the weak form, this scheme can also be formulated with respect to other configurations.

5.2.1 Tensile Instability

The OTM method is based on the use of local maximum entropy (LME) functions from Sect. 4.6.3. In the original form, the equations are formulated with respect to the updated Lagrangian configuration. Thus, a suitable approach is needed to avoid negative effects due to tensile instability. In Arroyo and Ortiz (2006) it is recommended to adapt the numerical parameter β of the LME function to the resolution of the computational domain

$$\beta = \frac{\gamma}{h^2}. \tag{5.17}$$

The parameter h correlates with the characteristic distance between nodes and corresponds to the element size in the FEM. The value γ characterizes the locality of the shape function. As the value increases, the influence of the nodes close to the edge of the support domain decreases. A too-large value can lead to a non-convergence of the algorithm for the determination of the unknown Lagrange multiplier. Rosolen et al. (2010) recommends a value in the range of $\gamma = 0.8 - 4$.

5.2.2 *Support Domain*

After updating the coordinates, the support domain for each integration point is redetermined. Two different approaches are presented.

Search algorithm. In Weißenfels and Wriggers (2018) a search algorithm is proposed to determine the nodes in the neighborhood of an integration point, see also Fig. 5.3. This approach is suitable for computing the fusion of particles in Additive Manufacturing processes. All nodes which distance is smaller than a previously defined radius $\alpha_p R_p$ corresponds to the set of nodes of the support domain p

$$H_p^{sup} = \left\{ \mathbf{x}_I \in B_t \middle| \|\mathbf{x}_I - \mathbf{x}_p\| < \alpha_p R_p \right\}. \tag{5.18}$$

The number of nodes in H_p^{sup} defines N_p^{sup}. The parameter α_p controls the size of the search area.

Update support domain. Another approach is to adapt the radius R of the support domain to the deformation of the body via the incremental deformation gradient

$$V_{p\,n+1} = \det \Delta\mathbf{F}_{p\,n+1} V_{p\,n}, \quad \rightarrow \quad R_{p\,n+1} = \left(\det \Delta\mathbf{F}_{p\,n+1}\right)^{1/3} R_{p\,n}. \tag{5.19}$$

With the redetermination of the search radius, the parameter β must be adjusted to reduce the negative effects due to a tensile instability. From (5.17), an update rule can be obtained for the 1-dimensional case. With (5.19) and replacing R with h this relation can be directly extended to the 3-dimensional case

$$\beta_{p\,n+1} = \frac{\gamma}{h_{p\,n+1}} = \beta_{p\,n} \left(\frac{h_{p\,n}}{h_{p\,n+1}}\right)^2 \quad \beta_{p\,n+1} = \frac{1}{\left(\Delta\mathbf{F}_{p\,n+1}\right)^{2/3}} \beta_{p\,n}. \tag{5.20}$$

1) Compute $r_p = \min\{\|\mathbf{x}_{I\,n+1} - \mathbf{x}_{p\,n+1}\|\}$ from new coordinates of the current support domain
2) Determine radius of new support domain $\alpha_p r_p$
3) Find nodes which belong to the integration point following (5.18)
4) If more than 4 nodes and if volume larger than volume of integration point
   Check all combinations until tetrahedron is found which encloses the integration point

   Else

   Increase factor $\alpha_p - \alpha_p * 1.2$ and go back to step 1)

Fig. 5.3 Algorithmic scheme to compute nodes which belong to the support domain of an integration point

The initial value R_0 can be estimated from the LME function. Due to the exponential character, the values decreases rapidly with increasing distance. With the definition of a small cutoff value ϵ, R_0 can be determined from (4.112) using the first formulation in (4.113)

$$\exp\left(-\beta R_0^2\right) = \epsilon \quad \rightarrow \quad R_0 = \sqrt{\frac{-\log \epsilon}{\beta}}. \tag{5.21}$$

This approach considers only the volumetric expansion of the support domain. However, the body can deform in any shape depending on the loading. To overcome this limitation, an anisotropic adjustment of the support domain is proposed in Kumar et al. (2019). Instead of a sphere, the nodes that lie inside an ellipsoid defines the support domain of the integration point p

$$\left(\mathbf{x}_{p\,n} - \mathbf{x}_{I\,n}\right) \mathbf{M}_{p\,n} \cdot \left(\mathbf{x}_{p\,n} - \mathbf{x}_{I\,n}\right) \leq 1. \tag{5.22}$$

The initialization of the matrix $\mathbf{M}$ can be realized using the assumption that the support domain is spherical at the beginning

$$\mathbf{M}_0 = \frac{1}{R_0^2}\mathbf{1}. \tag{5.23}$$

The redetermination of the search area results from the update of line elements

$$\mathbf{x}_{p\,n+1} - \mathbf{x}_{I\,n+1} = \Delta\mathbf{F}_{p\,n+1}\left(\mathbf{x}_{p\,n} - \mathbf{x}_{I\,n}\right). \tag{5.24}$$

The tensor $\mathbf{M}$ at the new time step follows directly from (5.22) using (5.24)

$$\mathbf{M}_{p\,n+1} = \Delta\mathbf{F}_{p\,n+1}^{-T}\mathbf{M}_{p\,n}\Delta\mathbf{F}_{p\,n+1}^{-1}. \tag{5.25}$$

Corresponding to the search area, the tensor $\boldsymbol{\beta}$ must also be adapted to avoid tensile instabilities. This quantity is updated in an analogous way to the tensor $\mathbf{M}$ using the second approach in (4.113) together with (5.24)

$$\boldsymbol{\beta}_{p\,n+1} = \Delta\mathbf{F}_{p\,n+1}^{-T}\boldsymbol{\beta}_{p\,n}\Delta\mathbf{F}_{p\,n+1}^{-1}. \tag{5.26}$$

5.2.3 Integration Constraint

The update of the support domain is not designed to meet all mandatory criteria of Sect. 4.5. The local maximum entropy (LME) functions fulfill the reproducing conditions for trial shape functions. They also possess a weak Kronecker-δ property. But standard search algorithms do not lead to the fulfillment of the integration constraint

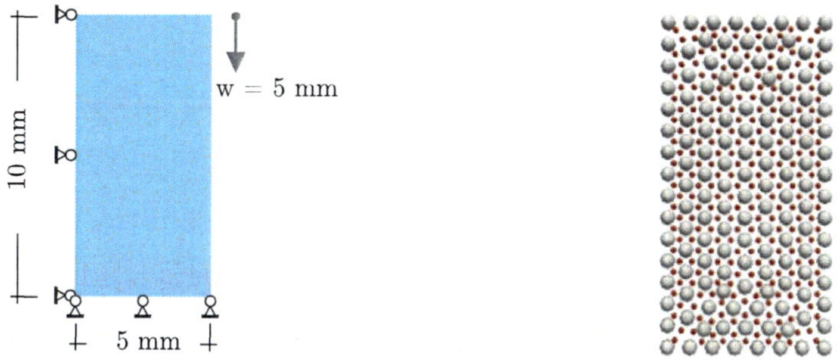

Fig. 5.4 (Left) Geometry, dimensions, and boundary conditions of the uniaxial test case. (Right) Subdivision of the domain into nodes (grey) and integration points (red)

automatically. The negative effects of such a violation can be shown using a simple test example. Due to the advantage regarding the simulation of fusion processes within Additive Manufacturing the search algorithm of Weißenfels and Wriggers (2018) is employed. A block is loaded at the upper side with a displacement of 5 mm in the vertical direction. Due to the uniaxial loading, negative effects due to under-integration are not expected. The geometry, dimensions, and boundary conditions can be found in Fig. 5.4. The material is assumed to behave non-linear elastic, and is approximated by a Neo–Hookean model. The Lame parameters can be derived from Young's modulus ($E = 1 \cdot 10^6$ N/mm^2) and Poisson ratio ($\nu = 0.2$)

$$\sigma_{p\,n+1} = \frac{\lambda}{2\,J_{p\,n+1}}\left(J^2_{p\,n+1} - 1\right) + \frac{\mu}{J_{p\,n+1}}\left(\mathbf{b}_{p\,n+1} - \mathbf{1}\right). \tag{5.27}$$

The load is applied in 10 steps. The parameter of the LME function is $\gamma = 2.0$ and the extension factor of the search algorithm is set initially to $\alpha = 1.5$. The true solution results in a homogeneous normal stress in vertical direction of $\sigma_{22} = -7.567 \cdot 10^5$ N/mm^2 and a horizontal displacement $u_x = 0.547$ mm. The equations are formulated with respect to the total Lagrangian description to avoid errors due to a violation of the configurational consistency and tensile instability. The OTM method without any stabilization or nodal shifting algorithm leads to an unphysical solution (Fig. 5.5). Since resulting nodal normal vectors also occur at inner nodes (Fig. 5.5), the integration constraint is violated and reasons the unphysical solution. To measure the violation the norm of the resulting normal vectors is summed up over all inner nodes n^{inner} and weighted by the volume of the body V

$$e^{IC} = \frac{1}{V}\sum_{I=1}^{n^{inner}} \|\mathbf{n}_I\|. \tag{5.28}$$

For the above example the error amounts to $e^{IC} = 0.196087$.

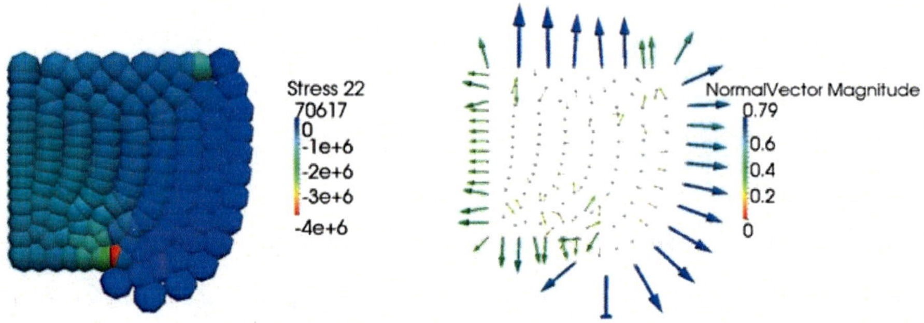

Fig. 5.5 Total Lagrangian OTM algorithm without stabilization or nodal shifting algorithm. (Left) Vertical normal nodal stress distribution σ_{22}. (Right) Resulting nodal normal vectors

Conclusion. The OTM method without any special treatment does not fulfill the integration constraint. Hence, an unphysical solution of the differential equation results. The integration constraint refers to the influence domain. The reproducing conditions of the trial shape function are based on the support domain. This complicates the simultaneous fulfillment of both requirements. Until now, there is no true meshfree Galerkin method that fulfills both criteria, if the integration is evaluated at separate points. A first formulation is presented in Sect. 5.2.6.

Remark In the original publication on the OTM method (Li et al. 2010) and in other publications, there is no information on the integration constraint. Only the need to use a nodal shifting algorithm based on r-adaptivity (Thoutireddy and Ortiz 2004; Mosler and Ortiz 2006) is mentioned briefly.

5.2.4 Configurational Consistency

Another reason for unphysical solutions can be the violation of the configurational consistency if the equations are formulated with respect to the updated Lagrangian description. To exclude negative effects from the search algorithm, the example from the previous section was calculated with a fixed assignment of nodes to integration points. Therefore, the domain is subdivided into 9-node elements (Fig. 5.6). The time step size is decreased to $\Delta t = 0.01$ s to emphasize possible negative effects more strongly. The outcome of the OTM method formulated with respect to the total Lagrangian description leads to the true solution since the integration constraint is fulfilled ($\sigma_{22} = -7.567 \cdot 10^5$ N/mm^2, u$_x$ = 0.547 mm). If the differential equation is given with respect to the updated Lagrangian configuration, an unphysical solution is obtained (Fig. 5.7). The negative effects result again from the violation of the integration constraint ($e^{IC} = 0.975286$). However, the non-fulfillment of the configurational consistency reasons the unphysical outcome. At each load step, the error is very small, but during the calculation this value accumulates (Fig. 5.6) which

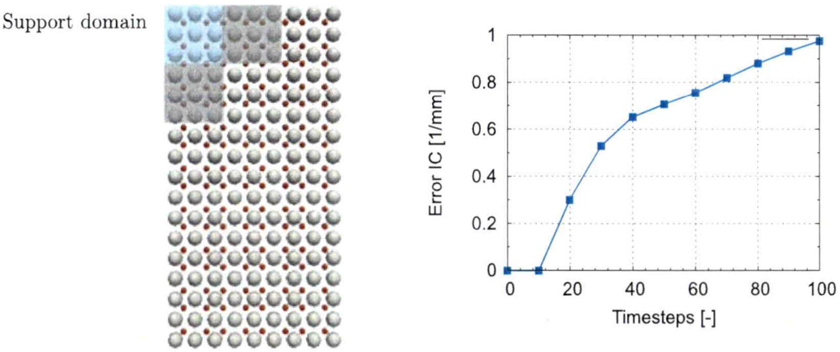

Fig. 5.6 (Left) Subdivision of the domain in non-overlapping support domains. (Right) Error due to a violation of the integration constraint over time using an updated Lagrangian description

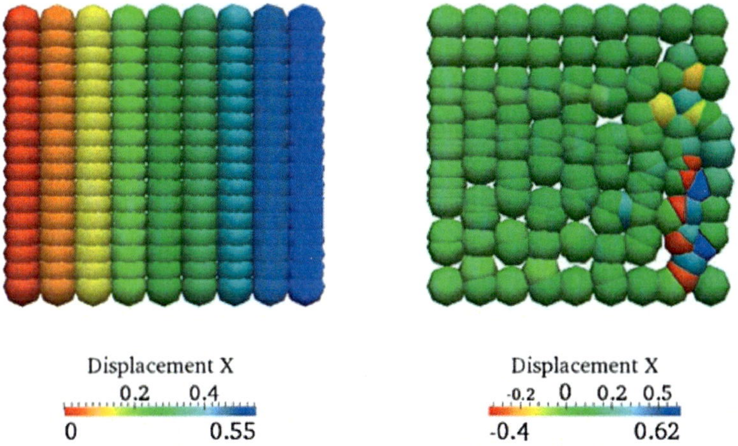

Fig. 5.7 Displacement boundary conditions: Horizontal nodal displacements. (Left) Total Lagrangian description. (Right) Updated Lagrangian description

can destroy the simulation results (Fig. 5.7). In Weißenfels and Wriggers (2018), a method was introduced which leads to better results using the OTM method. The motivation for this approach was the stabilization of under-integration. In fact, the system is forced to fulfill the configurational consistency using the penalty method. In this approach, a second contribution is added to the internal force. This extra term results from the multiplication of the normalized error of (4.84) with the penalty parameter ϵ

$$\mathbf{f}^{s,corr}_{I\,n+1} = \mathbf{f}^{s}_{I\,n+1} + \epsilon \sum_{I=1}^{N^{sup}_{p\,n}} N_I(\mathbf{x}_{p\,n}) \frac{\mathbf{x}_{I\,n+1} - \mathbf{x}_{p\,n+1} - \Delta\mathbf{F}_{p\,n+1}\left[\mathbf{x}_{I\,n} - \mathbf{x}_{p\,n}\right]}{\|\mathbf{x}_{I\,n} - \mathbf{x}_{p\,n}\|}. \tag{5.29}$$

To weight integration points closer to the node more strongly, the error is multiplied by the shape function. The application of this concept leads to accurate vertical

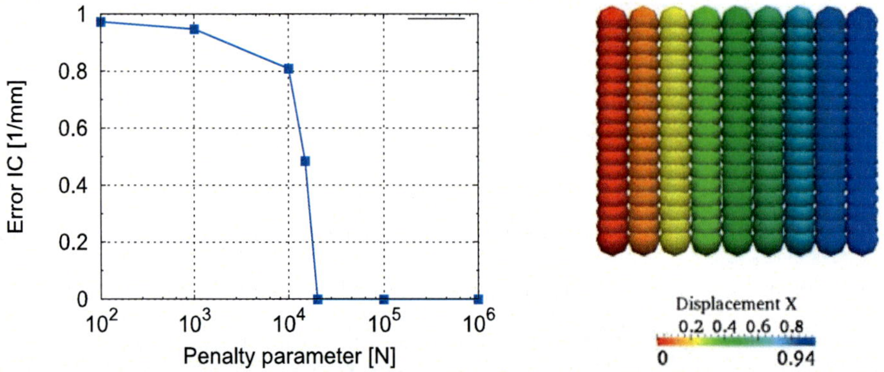

Fig. 5.8 Updated Lagrangian description. (Left) Influence of the penalty parameter on the error due to a violation of the integration constraint. (Right) Horizontal nodal displacements using a penalty parameter of $\epsilon = 10 \cdot 10^6$

stresses, if the penalty parameter is sufficiently high. In this case, the error due to the violation of the integration constraint is almost zero (Fig. 5.8). However, the horizontal displacements are larger compared to the true solution (Fig. 5.8).

When calculating the above example with a force boundary condition, the situation is even worse. To demonstrate the negative effects, a line load of $4 \cdot 10^4$ N/mm in vertical direction is applied on top of the block instead of the displacements. The formulation with respect to the total Lagrangian description leads again to the true solution (Fig. 5.9). If the updated Lagrangian formulation is employed a result can only be obtained with a penalty parameter larger than $\epsilon = 1 \cdot 10^5$. However, the vertical displacements are much stronger compared to the total Lagrangian formulation.

In the last test, the effects of increasing the number of integration points are examined. Again, the updated Lagrangian formulation is considered. The stabilization given in (5.29) is neglected. If displacement boundary conditions are applied, an increase in the total number of integration points from 128 to 288 shows a stabilized effect (Fig. 5.10). The true vertical stress σ_{22} results, but like in the stabilized case the horizontal displacements are over predicted. However, the error due to the violation of the integration constraint is almost zero ($e^{IC} = 0.00390164$). If a force boundary condition is imposed and the number of integration points is increased to 1152 a solution can be computed. However, like in the stabilized case, the vertical deformation is too strong. Additionally, an error due to the violation of the integration constraint is still present ($e^{IC} = 0.918779$).

Conclusion. If the OTM method is not formulated with respect to the total Lagrangian configuration, the configurational consistency can be violated which leads to inaccurate results. A correction scheme, which is often used in meshfree methods, enforces the algorithm to exactly reproduce the continuum-mechanical update of line elements. Using this stabilization the error due to a violation of the integration constraint can be corrected. However, unphysical results occur nevertheless. An increase of integration points leads to a similar unphysical outcome like in the stabilized case.

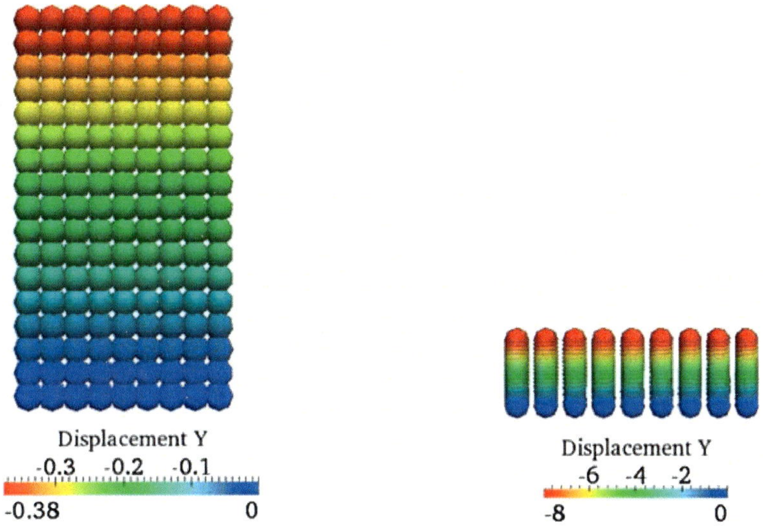

Fig. 5.9 Force boundary conditions: Vertical nodal displacements. (Left) Total Lagrangian description. (Right) Updated Lagrangian description using a stabilization with $\epsilon = 10 \cdot 10^5$

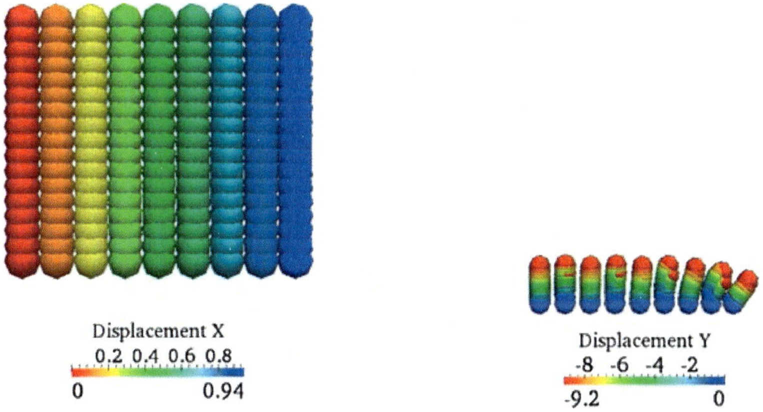

Fig. 5.10 (Left) Displacement boundary condition: Horizontal nodal displacements in the case of 288 integration points. (Right) Force boundary condition: Vertical nodal displacements in the case of 1152 integration points

5.2.5 Rank Instability

In the Finite Element Method, under-integration usually results in an unphysical solution, which can especially be observed under loadings leading to bending. The behavior of a cantilever arm loaded vertically is investigated to determine whether the OTM method shows negative effects due to under-integration. The line load at the right edge is set to 16 N/mm. The Lame parameters in (5.27) are chosen as $\lambda = 400943.27\,\text{N/mm}^2$ and $\mu = 80.194\,\text{N/mm}^2$. The load is applied in 50 steps.

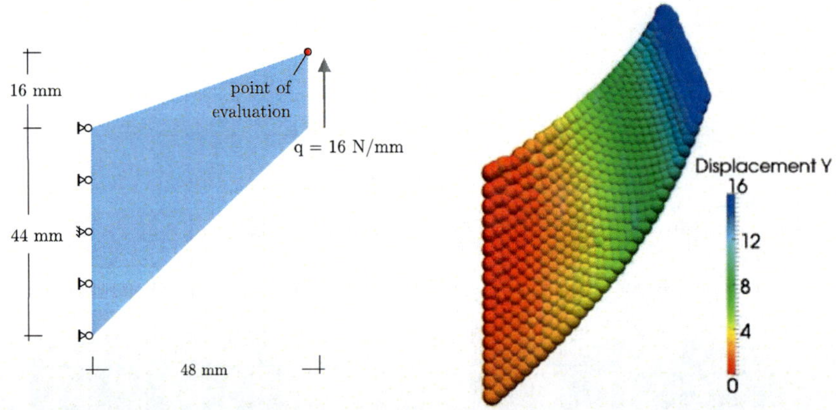

Fig. 5.11 (Left) Geometry, dimensions, and boundary conditions of the cantilever. (Right) Vertical nodal displacements using the OTM method in a Lagrangian description without stabilization

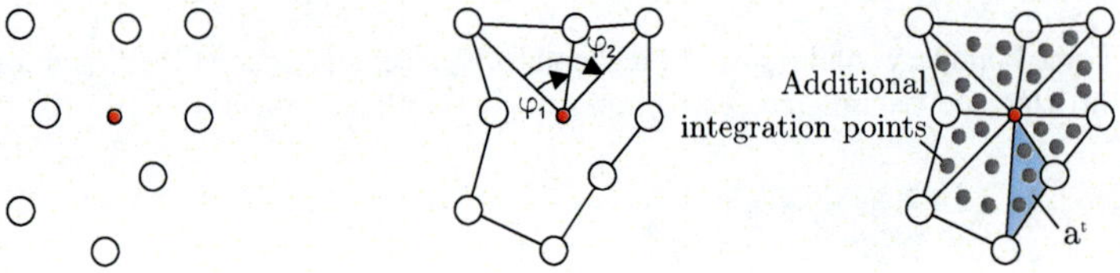

Fig. 5.12 Subdivision of support domain into triangles and localization of the corresponding integration points within each triangle

The geometry, dimensions, and boundary conditions can be found in Fig. 5.11. The search radius parameter is chosen so that effects from the violation of the integration constraint can be neglected, i.e. $\alpha = 4.5$. The parameter for the LME shape function is $\gamma = 2.0$. In addition, the differential equation is formulated with respect to the total Lagrangian description to exclude negative effects from the violation of the configurational consistency and tensile instability. A closer look at Fig. 5.11 shows a non-smooth displacement field. To investigate whether these unphysical variations result from under-integration, a concept for adding more evaluation points per support domain is implemented. In contrast to the previous example, the number of support domains remains constant. After the search algorithm, each domain is subdivided into triangles (Fig. 5.12). A different number of integration points can be located in each triangle

$$\mathbf{f}^s_{I\,n+1} = \sum_{t=1}^{N^{tri}_{p\,n+1}} \sum_{g=1}^{N^{gp}} \sum_{I=1}^{N^{sup}_{p\,n+1}} \boldsymbol{\sigma}_{g\,n+1} \frac{\partial N_I(\mathbf{X}_g)}{\partial \mathbf{x}_{g\,n+1}} v^t_{g\,n+1}. \tag{5.30}$$

The area $v^t_{n+1\,g}$ at the additional integration points is calculated from the total area $v_{p\,n+1}$ and the area of the corresponding triangle $a^t_{p\,n+1}$ using the Gauss quadrature rule

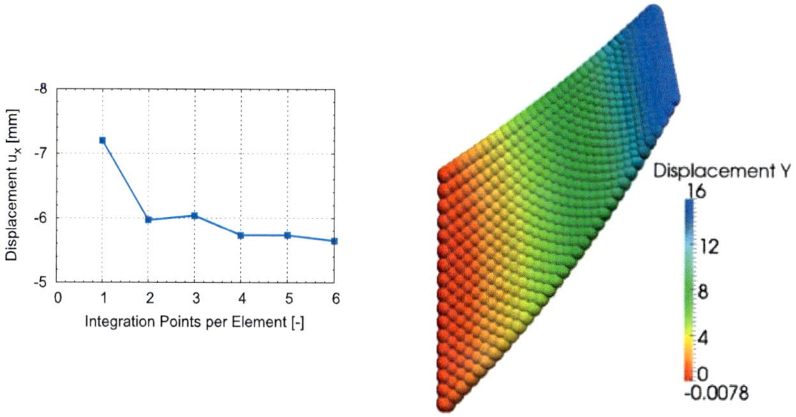

Fig. 5.13 (Left) Convergence behavior of horizontal displacement of upper right vertex by adding more integration points per triangle. (Right) Vertical nodal displacements in case of 1 integration point per triangle

$$v^t_{n+1\,g} = v_{p\,n+1} \frac{a^t_{p\,n+1}}{a^{sum}_{p\,n+1}} W_g, \qquad a^{sum}_{p\,n+1} = \sum_{i=1}^{N^{tri}_{p\,n+1}} a^i_{p\,n+1}. \tag{5.31}$$

The weights W_g for different integration points in a triangle can be found in e.g. Dhatt and Touzot (1984). As an example, the horizontal displacement at the upper right point (5.11) is evaluated. Only more than four integration points per triangle result in a convergent behavior (Fig. 5.13). The displacement field is also more smooth. On the other hand, the curvature is less (Fig. 5.13) compared to the original OTM formulation in Fig. 5.11. A convergence analysis shows a very stiff behavior of the formulation with additional integration points (Fig. 5.14). This effect also occurs with the FEM method when the domain is subdivided into 9-node elements and LME shape functions are used. This requires 64 integration points per element to accurately evaluate the integral. Adding the stabilization of (5.29), which was initially designed to correct negative effects from under-integration, the OTM formulation shows an improved behavior. However, the optimal penalty parameter depends on the number of degrees of freedom, as can be seen in Fig. 5.14 where only a constant value of $\epsilon = 2 \cdot 10^2$ is considered.

Conclusion. The OTM method solves the integral with a reduced number of integration points. Hence, effects known from under-integration occur, whereby the solution is not very strongly distorted. The application of LME functions leads to an excessively stiff behavior if sufficient integration points are introduced. This effect becomes especially visible when comparing with the FEM (Fig. 5.14). Hence, in the OTM method, under-integration reduces locking or excessively stiff behavior.

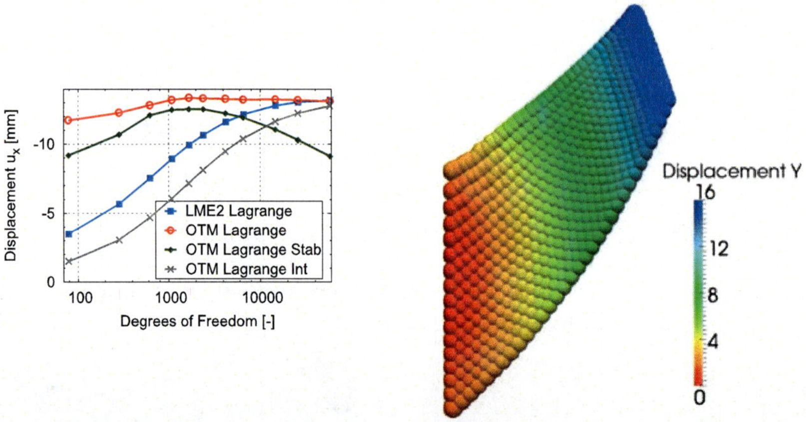

Fig. 5.14 (Left) Convergence plot of the horizontal displacement of the upper right vertex. (Right) Vertical nodal displacements in case of the stabilized OTM method

5.2.6 Petrov Galerkin OTM Method

The previous examples show that the integration constraint is one of the most important criteria for Galerkin methods. Therefore, an algorithm is developed that guarantees the fulfillment of this condition. The integration constraint refers to the test function in the influence domain. A Petrov–Galerkin formulation is chosen in order not to impair the reproducing conditions of the trial function in the support domain. In oder to avoid negative effects due to the violation of the configurational consistency or tensile instability the equations are formulated with respect to the initial configuration. The discretization of the displacements is analogous to Sect. 4.6.3. For the test function, a modified LME approach is used which is now related to the influence domain

$$\begin{aligned} N_I^\eta\left(\mathbf{X}_p\right) &= \frac{Z_I^\eta\left(\mathbf{X}_p\right)}{Z^\eta}, \quad Z^\eta = \sum_{p=1}^{N_I^{inf}} Z_I^\eta\left(\mathbf{X}_p\right), \\ Z_I^\eta\left(\mathbf{X}_p\right) &= \exp\left(-\beta\|\mathbf{X}_p - \mathbf{X}_I\|\right)\exp\left(\boldsymbol{\lambda}_I \cdot \left[\mathbf{X}_p - \mathbf{X}_I\right]\right). \end{aligned} \tag{5.32}$$

The Lagrange multiplier is defined at the node and is calculated by fulfilling the integration constraint

$$\mathbf{r}_I\left(\mathbf{X}_p, \boldsymbol{\lambda}_p\right) = \sum_{p=1}^{N_I^{inf}} \frac{\partial N_I(\mathbf{X}_p)}{\partial \mathbf{X}_p} v_{0\,p} = \begin{cases} 0 & \text{for inner nodes} \\ \bar{\mathbf{N}}_I & \text{for boundary nodes} \end{cases}. \tag{5.33}$$

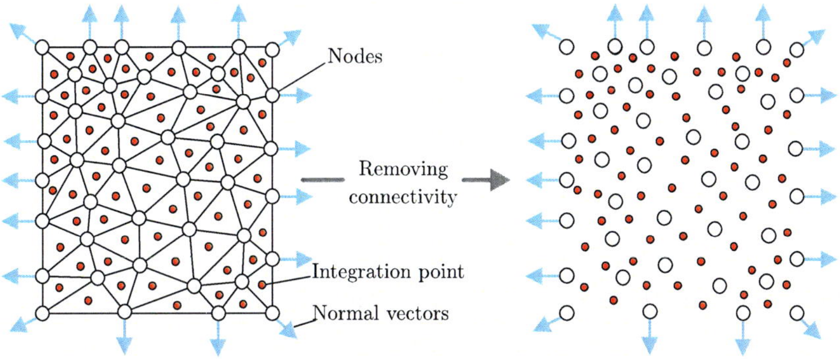

Fig. 5.15 Initial subdivision of the domain into triangles in 2D and determination of normal vectors. Removal of mesh results in the location of integration points and nodes as well as normal vectors

In this equation, the resulting normal vector at the boundary nodes is requested. These vectors must be known at the beginning of the calculation. The initialization of the OTM method is based on a subdivision of the domain into elements. As shown in Weißenfels (2019), the Finite Element Method fulfills the integration constraint. Hence, the resulting normal vector at the boundary node $\bar{\mathbf{N}}_I$ from an FEM discretization can be used for the initialization of the normal vectors (Fig. 5.15). During the simulation the current vector $\bar{\mathbf{n}}_I$ results from an update using the deformation gradient

$$\bar{\mathbf{n}}_I = \sum_{p=1}^{N_I^{inf}} \frac{N_I(\mathbf{X}_p)}{\partial \mathbf{X}_p} \mathbf{F}_p^{-1} \det \mathbf{F}_p v_{p\,0}. \tag{5.34}$$

The above equation corresponds to the discrete form of the Nanson formula, which relates surface integrals between the initial and the current configuration. The Nanson formula can be found e.g. in Wriggers (2008). Analogous to the standard LME shape functions, (5.33) is nonlinear with respect to the unknown Lagrange multiplier $\boldsymbol{\lambda}_I$. This vector is calculated using the Newton-Raphson iteration. To simplify the implementation the program package AceGen (Korelc 1997; Korelc and Wriggers 2016) can be used to automatically calculate the tangent matrix.

The test from Sect. 5.2.3 is conducted to examine the significance of the integration constraint. The geometry, dimensions, and boundary conditions can be found in Fig. 5.4. The original OTM method leads to a violation of the integration constraint and inner nodes possess a resulting normal vector that is not equal to zero (Fig. 5.5). This violation leads to an unphysical deformation of the body. If the Petrov–Galerkin variant is chosen that automatically satisfies the integration constraint of 0th order, the resulting normal vectors inside are identical zero and the true solution is obtained (Fig. 5.16). This example highlights the importance of the integration constraint for fulfilling the patch test.

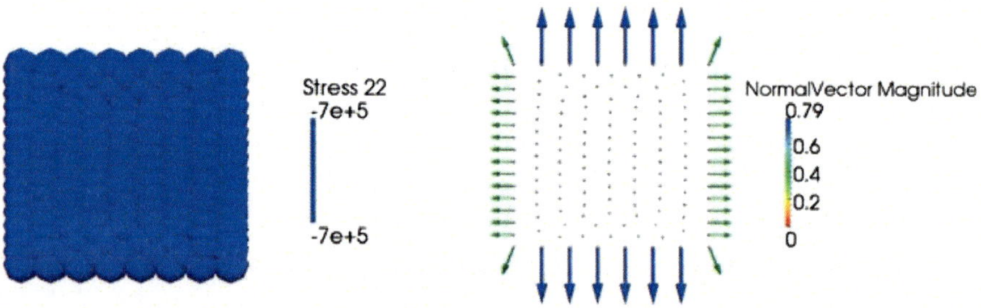

Fig. 5.16 Lagrangian Petrov–Galerkin OTM algorithm (Left) Vertical normal nodal stress (σ_{22}) distribution. (Right) Resulting nodal normal vectors

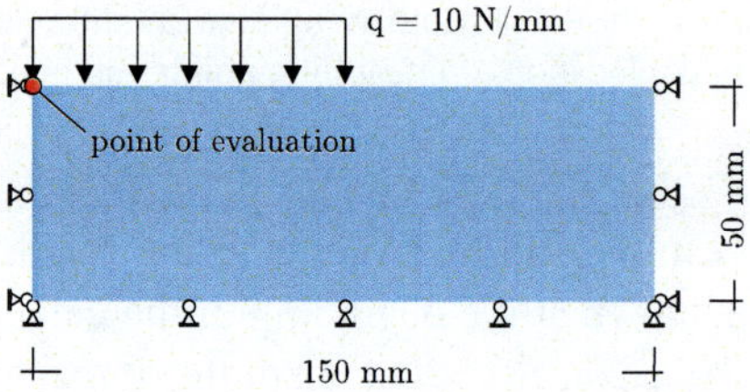

Fig. 5.17 Geometry, dimensions, and boundary conditions of the punch test

In the second example, the half of the upper surface of a block is loaded vertically with a line load of 10 N/mm. The geometry, dimensions, and boundary conditions can be found in Fig. 5.17. If the force at the nodes is assumed to be constant, necking in the middle of the upper block can be observed. This necking results from the change of the upper surface at strong deformations. If the forces are assumed to be constant, the stress in the middle is larger than at the vertex. Thus, the load no longer corresponds to a constant line load. To avoid this negative effect, the stress at the upper boundary is defined as a follower load

$$\mathbf{f}_I^{ext} := q\,\mathbf{n}_I, \qquad \mathbf{n}_I = \sum_{p=1}^{N_I^{inf}} \frac{N_I(\mathbf{X}_p)}{\partial \mathbf{X}_p} \mathbf{F}_p^{-1} \det \mathbf{F}_p v_{p\,0}. \tag{5.35}$$

According to the approach from Weißenfels (2019), the current normal vector and the nodal area result directly from the integration constraint (5.34). The load is applied in 50 steps. The material is defined as non-linear elastic according to (5.27). The Young's modulus is selected to $E = 7\,\text{N/mm}^2$ and the Poisson ratio to $\nu = 0.44$. The parameter for the LME function is $\gamma = 2.0$. The value for the search radius is selected as $\alpha = 2.5$. The displacement field of the Petrov–Galerkin OTM method corresponds to the FEM solution employing linear triangular elements (Fig. 5.18). In addition, the convergence behavior shows that the algorithm is very stiff and does not converge smoothly (Fig. 5.18).

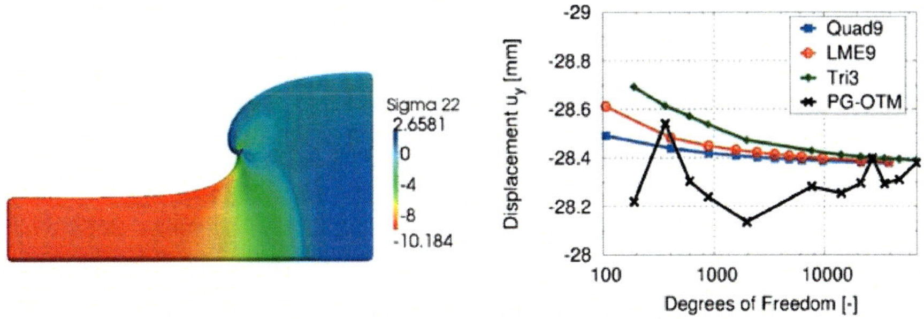

Fig. 5.18 Punch test: (Left) Vertical normal nodal stress (σ_{22}) distribution. (Right) Convergence diagram comparing FEM solutions using 9-node polynomial (Quad9), 9-node LME (LME9), and 3-node polynomial (Tri3) shape functions with the outcome of the PG-OTM approach

Conclusion. The fulfillment of the integration constraint is a prerequisite for an accurate solution of the differential equation. Using a Petrov–Galerkin formulation, the OTM method can be forced to fulfill this condition. On the other hand, this variant leads to a stiff response. This behavior can also be derived from the studies in the previous section. Therein, it is shown that LME shape functions favor a too stiff response behavior. The outcomes are similar to a FEM solution with linear triangular elements. In addition, the algorithm is not stable for weakly compressible material behavior.

Remark 1 Using a Bubnov–Galerkin approach the test N_I^η and the trial N_I^u functions are identical. This leads in the case of a conservative system, like in the previous examples, to a symmetric tangent that is physically reasonable. The Petrov–Galerkin formulation can also be interpreted as a disturbed Bubnov–Galerkin approach since each test function can be decomposed into a trial function and a disturbed contribution $\tilde{N}_I$

$$N_I^\eta = N_I^u + \tilde{N}_I. \tag{5.36}$$

The second term leads to an additional contribution and reasons the unsymmetric tangent. However, the additional term corrects the shortcomings of the Bubnov–Galerkin approach.

Remark 2 The imposition of a surface load as follower load, like in the last example, also leads to an unsymmetric tangent. In this case, one speaks of a body-attached follower load. This is unphysical since static loads are conservative and must, therefore, lead to a symmetric tangent (Schweizerhof and Ramm 1984). However, in the last example, physically reasonable results are obtained.

References

M. Arroyo, M. Ortiz, Local maximum-entropy approximation schemes: a seamless bridge between finite elements and meshfree methods. Int. J. Numer. Meth. Eng. **65**, 2167–2202 (2006)

S. Beissel, T. Belytschko, Nodal integration of the element-free Galerkin method. Comput. Methods Appl. Mech. Eng. **139**(1), 49–74 (1996)

T. Belytschko, Y.Y. Lu, L. Gu, Element-free Galerkin methods. Int. J. Numer. Meth. Eng. **37**, 229–256 (1994)

T. Bode, C. Weißenfels, P. Wriggers, A consistent peridynamic formulation for arbitrary particle distributions. Comput. Methods Appl. Mech. Eng. **374**, 113605 (2021)

J. Bonet, S.D. Kulasegaram, Correction and stabilization of Smooth Particle Hydrodynamics methods with applications in metal forming simulations. Int. J. Numer. Meth. Eng. **47**(6), 1189–1214 (2000)

J.S. Chen, C.T. Wu, S. Yoon, Y. You, A stabilized conforming nodal integration for Galerkin meshfree methods. Int. J. Numer. Meth. Eng. **50**(2), 435–466 (2001)

J.S. Chen, S. Yoon, C.T. Wu, Non-linear version of stabilized conforming nodal integration for Galerkin mesh-free methods. Int. J. Numer. Meth. Eng. **53**(12), 2587–2615 (2002)

J.S. Chen, W. Hu, M.A. Puso, Y. Wu, X. Zhang, Strain smoothing for stabilization and regularization of Galerkin meshfree methods, in *Meshfree Methods for Partial Differential Equations III*, pp. 57–75 (Springer, Berlin, 2007)

J.S. Chen, M. Hillman, M. Rüter, An arbitrary order variationally consistent integration for Galerkin meshfree methods. Int. J. Numer. Meth. Eng. **95**(5), 387–418 (2013)

S. De, K.J. Bathe, The method of finite spheres with improved numerical integration. Comput. Struct. **79**(22), 2183–2196 (2001)

G. Dhatt, G. Touzot, *The Finite Elemet Method Displayed* (Wiley, Chicester, 1984)

J. Dolbow, T. Belytschko, Numerical integration of the Galerkin weak form in meshfree methods. Comput. Mech. **23**(3), 219–230 (1999)

M. Hillman, J.S. Chen, An accelerated, convergent, and stable nodal integration in galerkin meshfree methods for linear and nonlinear mechanics. Int. J. Numer. Meth. Eng. **107**(7), 603–630 (2016)

J. Korelc, Automatic generation of finite-element code by simultaneous optimization of expressions. Theoret. Comput. Sci. **187**, 231–248 (1997)

J. Korelc, P. Wriggers, *Automation of Finite Element Methods* (Springer, Berlin, 2016)

S. Kumar, K. Danas, D.M. Kochmann, Enhanced local maximum-entropy approximation for stable meshfree simulations. Comput. Methods Appl. Mech. Eng. **344**, 858–886 (2019)

B. Li, F. Habbal, M. Ortiz, Optimal transportation meshfree approximation schemes for fluid and plastic flows. Int. J. Numer. Meth. Eng. **83**, 1541–1579 (2010)

S. Li, W.K. Liu, *Meshfree Particle Methods* (Springer, Berlin, 2007)

W.K. Liu, S. Jun, Y.F. Zhang, Reproducing kernel particle methods. Int. J. Numer. Meth. Fluids **20**(8–9), 1081–1106 (1995)

W.K. Liu, S. Li, T. Belytschko, Moving least-square reproducing kernel methods (I): methodology and convergence. Comput. Methods Appl. Mech. Eng. **143**(1), 113–154 (1997)

J. Mosler, M. Ortiz, On the numerical implementation of variational arbitrary Lagrangian-Eulerian (VALE) formulations. Int. J. Numer. Meth. Eng. **67**(9), 1272–1289 (2006)

B. Nayroles, G. Touzot, P. Villon, Generalizing the finite element method: diffuse approximation and diffuse elements. Comput. Mech. **10**(5), 307–318 (1992)

M.A. Puso, J.S. Chen, E. Zywicz, W. Elmer, Meshfree and finite element nodal integration methods. Int. J. Numer. Meth. Eng. **74**(3), 416–446 (2008)

A. Rosolen, D. Millán, M. Arroyo, On the optimum support size in meshfree methods: a variational adaptivity approach with maximum-entropy approximants. Int. J. Numer. Meth. Eng. **82**(7), 868–895 (2010)

K. Schweizerhof, E. Ramm, Displacement dependent pressure loads in nonlinear finite element analysis. Comput. & Struct. **6**, 1099–1114 (1984)

P. Thoutireddy, M. Ortiz, A variational r-adaption and shape-optimization method for finite-deformation elasticity. Int. J. Numer. Meth. Eng. **61**(1), 1–21 (2004)
C. Villani, *Topics in Optimal Transportation Theory*, Graduate Studies in Mathematics, vol. 58, 2nd edn. (American Mathematical Society, Providence, 2013)
C. Weißenfels, Direct nodal imposition of surface loads using the divergence theorem. Finite Elem. Anal. Des. **165**, 31–40 (2019)
C. Weißenfels, P. Wriggers, Stabilization algorithm for the optimal transportation meshfree approximation scheme. Comput. Methods Appl. Mech. Eng. **329**, 421–443 (2018)
P. Wriggers, *Nonlinear Finite Element Methods* (Springer Science & Business Media, Berlin, 2008)

Chapter 6
Smoothed Particle Hydrodynamics

Smoothed Particle Hydrodynamics (SPH) has experienced a renaissance in computational mechanics in recent years. Due to the increased computing power, the additional effort due to the search algorithm, among other things, is no longer so significant. Nevertheless, some shortcomings are still present as will be shown in this chapter.

Smoothed Particle Hydrodynamics (SPH) is a spatial discretization scheme developed in the seventies by Gingold and Monaghan (1977) and in parallel by Lucy (1977). The first applications were limited to the field of astrophysics. In the meantime, this method has also become established in computational mechanics and is used especially for the simulation of fluid flows with free surfaces. Although the formulation is equally suitable for fluids and solids, different variants have been established for the two scenarios.

The SPH subdivides the domain into individual particles at which all quantities are evaluated. In the SPH, own technical terms have been established over the years. In the individual sections, the corresponding definitions are discussed and related to the usual expressions from Chap. 4.

6.1 Discretization Concept

The starting point is the description of the approximate solution in terms of an integral function similar to Sect. 4.6.4. To be able to express $\mathbf{u}$ at any point, the kernel function must possess a reproducing property. This is given for the δ-Dirac function

$$\mathbf{u}(\mathbf{x}) = \int_{H_{\mathbf{x}}} \delta\left(\mathbf{x} - \mathbf{x}'\right) \mathbf{u}\left(\mathbf{x}'\right) \, dv_{\mathbf{x}'} . \tag{6.1}$$

On the other hand, the δ-Dirac function is neither continuous nor differentiable. Due to these pathological properties, this approach is not suitable in the context of

C. Weißenfels, *Simulation of Additive Manufacturing using Meshfree Methods*,
Lecture Notes in Applied and Computational Mechanics 97,
https://doi.org/10.1007/978-3-030-87337-0_6

a discretization (Li and Liu 2007). In SPH, an attempt is made to get as close as possible to the reproducing properties by means of a smooth differentiable function, the so-called weighting or kernel function W. The values of W in the domain can be controlled by the so-called smoothing length h

$$\langle \mathbf{u}(\mathbf{x}) \rangle = \int_{H_\mathbf{x}} W\left(\mathbf{x} - \mathbf{x}', h\right) \mathbf{u}\left(\mathbf{x}'\right) \, \mathrm{d}v_{\mathbf{x}'}. \tag{6.2}$$

Since W is mostly chosen such that it violates the reproducing property, the result does not match the function at the corresponding position. In the equation above this is indicated by the angle bracket. Rather, (6.2) can be understood as an averaging. Alternatively, the value on the right-hand side can be changed. Thus, the function $\mathbf{u}$ can be approximated by the unknown values $\mathbf{a}$ in the neighborhood

$$\mathbf{u}(\mathbf{x}) = \int_{H_\mathbf{x}} W\left(\mathbf{x} - \mathbf{x}', h\right) \mathbf{a}\left(\mathbf{x}'\right) \, \mathrm{d}v_{\mathbf{x}'}. \tag{6.3}$$

There are two requirements on the weighting function in SPH. In the limit as the smoothing length approaches zero W must equal the δ-Dirac function. In addition, to ensure that for a constant displacement field the left-hand side in (6.2) reflects the correct value of $\mathbf{u}$, a normalization property is required

$$\lim_{h \to 0} W\left(\mathbf{x} - \mathbf{x}', h\right) = \delta\left(\mathbf{x} - \mathbf{x}'\right), \quad \int_{H_\mathbf{x}} W\left(\mathbf{x} - \mathbf{x}', h\right) \, \mathrm{d}v_{\mathbf{x}'} = 1. \tag{6.4}$$

In the discrete form, the sum over all particles $N_\mathbf{x}$ in the neighborhood of $\mathbf{x}$ replaces the integral in (6.3)

$$\mathbf{u}(\mathbf{x}) = \sum_{J=1}^{N_\mathbf{x}} W\left(\mathbf{x} - \mathbf{x}_J, h\right) \mathbf{a}_J v_J. \tag{6.5}$$

The unit of W is $1/\mathrm{m}^3$. The multiplication of the weighting function by the volume can be interpreted as a shape function

$$\mathbf{u}(\mathbf{x}) = \sum_{J=1}^{N_\mathbf{x}} N_J(\mathbf{x}) \mathbf{a}_J, \qquad N_J(\mathbf{x}) = W(\mathbf{x} - \mathbf{x}_J, h) v_J. \tag{6.6}$$

The weighting function is usually based on the distance $r = \|\mathbf{x} - \mathbf{x}_J\|$. The value has its maximum at $\mathbf{x}$ and decreases towards the boundary of the neighborhood. The size of the smoothing length h usually correlates with the radius of the search algorithm (Li and Liu 2007). To simplify the notation, instead of explicitly specifying the coordinates, only the index of the particle is given, i.e.

$$v_I := v\left(\mathbf{x}_I\right), \quad \mathbf{f}_I^s := \mathbf{f}^s\left(\mathbf{x}_I\right), \quad \boldsymbol{\sigma}_I := \boldsymbol{\sigma}\left(\mathbf{x}_I\right). \tag{6.7}$$

In addition, different abbreviations exist for the weighting function

$$W(\mathbf{x}_J - \mathbf{x}_I, h) := W(\mathbf{x}_J - \mathbf{x}_I) = W_I(\mathbf{x}_J) = W_{IJ}. \tag{6.8}$$

For the calculation of derivatives, such as the divergence of stress, the integral from (6.2) has to be differentiated

$$\langle \operatorname{div} \boldsymbol{\sigma}(\mathbf{x}) \rangle = \frac{\partial}{\partial \mathbf{x}} \int_{H_\mathbf{x}} W\left(\mathbf{x} - \mathbf{x}'\right) \boldsymbol{\sigma}\left(\mathbf{x}'\right) \, \mathrm{d}v_{\mathbf{x}'} : \mathbf{1} = \int_{H_\mathbf{x}} \boldsymbol{\sigma}\left(\mathbf{x}'\right) \frac{\partial W\left(\mathbf{x} - \mathbf{x}'\right)}{\partial \mathbf{x}} \, \mathrm{d}v_{\mathbf{x}'}. \tag{6.9}$$

The operator : denotes the double contraction between two tensors. The use of radial weighting functions leads to properties which can be advantageous in the context of a discretization. Since W is based on the distance between two points, the arguments of the function can be interchanged

$$W\left(\mathbf{x} - \mathbf{x}'\right) = W\left(\mathbf{x}' - \mathbf{x}\right). \tag{6.10}$$

For the derivative of the weighting function this property results in quite useful relations

$$\frac{\partial W\left(\mathbf{x} - \mathbf{x}'\right)}{\partial \mathbf{x}} = -\frac{\partial W\left(\mathbf{x} - \mathbf{x}'\right)}{\partial \mathbf{x}'} = -\frac{\partial W\left(\mathbf{x}' - \mathbf{x}\right)}{\partial \mathbf{x}'} = \frac{\partial W\left(\mathbf{x}' - \mathbf{x}\right)}{\partial \mathbf{x}}. \tag{6.11}$$

The same applies for the discrete case

$$\frac{\partial W\left(\mathbf{x}_I - \mathbf{x}_J\right)}{\partial \mathbf{x}_I} = -\frac{\partial W\left(\mathbf{x}_I - \mathbf{x}_J\right)}{\partial \mathbf{x}_J} = -\frac{\partial W\left(\mathbf{x}_J - \mathbf{x}_I\right)}{\partial \mathbf{x}_J} = \frac{\partial W\left(\mathbf{x}_J - \mathbf{x}_I\right)}{\partial \mathbf{x}_I}. \tag{6.12}$$

A more detailed introduction to the discretization concept of SPH can be found, for example, in Li and Liu (2007) or Liu and Liu (2003).

Remark 6.1 Due to the dimension, the weighting function can also be assumed to be an operator that determines the density at a point from the mass in the neighborhood (Price 2012)

$$\rho(\mathbf{x}) = \sum_{J=1}^{N_\mathbf{x}} W(\mathbf{x} - \mathbf{x}_J, h) m_J. \tag{6.13}$$

If the mass m_J at each particle is constant during the calculation, the conservation of mass is always satisfied.

Remark 6.2 In Monaghan (1992), so-called golden rules for a better approximation are defined, without giving an exact justification. Therein, it is recommended to use a Gaussian weighting function. However, mostly Wendland functions (Wendland 1995) are chosen for W, e.g.

$$W(\mathbf{x}-\mathbf{x}_J,h)=\frac{1}{\pi h^3}\begin{cases}\left(1-\frac{r}{2h}\right)^4\left(1+4\frac{r}{2h}\right) & \text{for } r\le 2h\\ 0 & \text{for } r>2h\end{cases}. \tag{6.14}$$

To ensure the correct unit, the function is divided by the volume. The distances are normalized by the smoothing length h.

6.2 Equivalence Weak Form and Reduced Order Strong Form

The discretization concept of SPH can be applied to all variants from Sect. 4.2. Normally, the domain is only subdivided into particles. An exception is the Normalized SPH (Randles and Libersky 2000), which is based on a two-point approach. Calculations of fluid flow are usually based on the strong form whereas simulations of solid deformations rely on the strong form with reduced order.

Substituting the relation between weighting and shape functions (6.6), the internal force of the weak form (4.14) can also be formulated as a function of W

$$\mathbf{f}_I^{s\,wf}=\sum_{J=1}^{N_I}\boldsymbol{\sigma}_J\frac{\partial N_I(\mathbf{x}_J)}{\partial \mathbf{x}_J}v_J=\sum_{J=1}^{N_I}\boldsymbol{\sigma}_J\frac{\partial W(\mathbf{x}_J-\mathbf{x}_I)}{\partial \mathbf{x}_J}v_I v_J. \tag{6.15}$$

Also, in the case of the reduced order strong form (4.23), this contribution can be specified in more detail

$$\mathbf{f}_I^{s\,ro}=-v_I\sum_{J=1}^{N_I}\boldsymbol{\sigma}_J\frac{\partial N_J(\mathbf{x}_I)}{\partial \mathbf{x}_I}=-\sum_{J=1}^{N_I}\boldsymbol{\sigma}_J\frac{\partial W(\mathbf{x}_I-\mathbf{x}_J)}{\partial \mathbf{x}_I}v_I v_J. \tag{6.16}$$

Exploiting the properties of radial functions (6.12), it can be shown that both approaches are equivalent

$$\mathbf{f}_I^{s\,ro}=-\sum_{J=1}^{N_I}\boldsymbol{\sigma}_J\frac{\partial W(\mathbf{x}_I-\mathbf{x}_J)}{\partial \mathbf{x}_I}v_I v_J=\sum_{J=1}^{N_I}\boldsymbol{\sigma}_J\frac{\partial W(\mathbf{x}_J-\mathbf{x}_I)}{\partial \mathbf{x}_J}v_I v_J=\mathbf{f}_I^{s\,wf}. \tag{6.17}$$

The contributions from the inertia term and the dead load are identical. This shows the equivalence between the weak form and the reduced order strong form in the framework of the SPH. Thus, even for the strong form, the forces due to the Neumann boundary conditions can be determined directly from (4.17).

Remark. In Randles and Libersky (1996), Belytschko et al. (1998) or Rabczuk et al. (2004) an alternative approach is used to calculate the derivative. This variant results directly from (6.9) taking into account (6.11)

$$\langle \operatorname{div} \boldsymbol{\sigma}(\mathbf{x}) \rangle = \int_{H_{\mathbf{x}}} \boldsymbol{\sigma}(\mathbf{x}') \frac{\partial W(\mathbf{x}-\mathbf{x}')}{\partial \mathbf{x}} \, dv_{\mathbf{x}'} = -\int_{H_{\mathbf{x}}} \boldsymbol{\sigma}(\mathbf{x}') \frac{\partial W(\mathbf{x}-\mathbf{x}')}{\partial \mathbf{x}'} \, dv_{\mathbf{x}'}. \tag{6.18}$$

From the equivalence theorem, the first relation corresponds to the derivative rule in the framework of the strong form and the second to the rule of the weak form.

6.3 Integration Constraint

In Bonet and Kulasegaram (2000), it is shown that even in the SPH an additional requirement must be fulfilled beside the reproducing conditions to satisfy the patch test. This requirement corresponds exactly to the integration constraint. One reason provides directly Sect. 6.2. Another justification results from the discretization concept of the SPH, which follows from an integral. Thus, the divergence of stress can alternatively be determined from the individual contributions in the neighborhood

$$\langle \operatorname{div} \boldsymbol{\sigma}(\mathbf{x}) \rangle = \int_{H_{\mathbf{x}}} W(\mathbf{x}-\mathbf{x}') \operatorname{div} \boldsymbol{\sigma}(\mathbf{x}') \, dv_{\mathbf{x}'}. \tag{6.19}$$

Using Gauss' integral theorem, the right-hand side can be related to the derivative of the weighting function

$$\begin{aligned} \int_{H_{\mathbf{x}}} W(\mathbf{x}-\mathbf{x}') \operatorname{div} \boldsymbol{\sigma}(\mathbf{x}') \, dv_{\mathbf{x}'} = & \int_{\partial H_{\mathbf{x}}} W(\mathbf{x}-\mathbf{x}') \boldsymbol{\sigma}(\mathbf{x}') \mathbf{n}(\mathbf{x}') \, da_{\mathbf{x}'} \\ & - \int_{H_{\mathbf{x}}} \boldsymbol{\sigma}(\mathbf{x}') \frac{\partial W(\mathbf{x}-\mathbf{x}')}{\partial \mathbf{x}'} \, dv_{\mathbf{x}'} \end{aligned} \tag{6.20}$$

Exploiting (6.10) and (6.11) as well as the transition $\mathbf{x}' \rightarrow \mathbf{x}$, $\mathbf{x} \rightarrow \mathbf{x}_I$ an alternative expression can be obtained

$$\begin{aligned} \int_{H_I} W(\mathbf{x}-\mathbf{x}_I) \operatorname{div} \boldsymbol{\sigma}(\mathbf{x}) \, dv_{\mathbf{x}} = & \int_{\partial H_I} W(\mathbf{x}-\mathbf{x}_I) \boldsymbol{\sigma}(\mathbf{x}) \mathbf{n}(\mathbf{x}) \, da_{\mathbf{x}} \\ & - \int_{H_I} \boldsymbol{\sigma}(\mathbf{x}) \frac{\partial W(\mathbf{x}-\mathbf{x}_I)}{\partial \mathbf{x}} \, dv_{\mathbf{x}}. \end{aligned} \tag{6.21}$$

Since $W(\mathbf{x}-\mathbf{x}_I) \, v_I = N_I(\mathbf{x})$, this relation corresponds exactly to the integration constraint given in (4.74). The standard patch test is satisfied if Eq. (6.20) is fulfilled for a constant stress

$$\int_{H_{\mathbf{x}}} \frac{\partial W(\mathbf{x}-\mathbf{x}')}{\partial \mathbf{x}'} \, dv_{\mathbf{x}'} = \int_{\partial H_{\mathbf{x}}} W(\mathbf{x}-\mathbf{x}') \mathbf{n}(\mathbf{x}') \, da_{\mathbf{x}'}. \tag{6.22}$$

In discrete form, this requirement results in a constraint that each particle must satisfy separately

$$v_I \sum_{J=1}^{N_I} \frac{\partial W(\mathbf{x}_I - \mathbf{x}_J)}{\partial \mathbf{x}_J} v_J = \begin{cases} 0 & \text{inner particles} \\ \bar{\mathbf{n}}_I & \text{boundary particles} \end{cases} . \tag{6.23}$$

The resulting normal vector at the boundary particle follows directly from the discretization of the right-hand side in (6.22)

$$\bar{\mathbf{n}}_I = \sum_{J=1}^{N_I} W(\mathbf{x}_I - \mathbf{x}_J)\, \mathbf{n}_J a_J . \tag{6.24}$$

If the condition from the integration constraint is violated, a correction is needed. An approach can be found in Bonet and Kulasegaram (2000) which is specified in more detail in Sect. 5.1.3.

Remark. In Belytschko et al. (1998) there is a hint that (6.20) can be a reason for an error in the approximation. However, the reference to the integration constraint is missing. Instead, the analysis bears upon the derivative rule according to (6.18).

6.4 Alternative Derivative Rules

Besides to the variants (6.9) and (6.18) there are two additional concepts in SPH to calculate the derivative of a quantity.

6.4.1 Product Rule

With the help of another quantity, such as the density, the divergence of stress can be reformulated

$$\operatorname{div}(\rho\boldsymbol{\sigma}) = \rho \operatorname{div}\boldsymbol{\sigma} + \boldsymbol{\sigma}\operatorname{grad}\rho \quad \text{leading to} \quad \operatorname{div}\boldsymbol{\sigma} = \frac{1}{\rho}\left[\operatorname{div}(\rho\boldsymbol{\sigma}) - \boldsymbol{\sigma}\operatorname{grad}\rho\right] . \tag{6.25}$$

Using the derivative rule (6.9), the divergence in SPH can also be calculated by means of a difference

$$\langle \operatorname{div}\boldsymbol{\sigma}(\mathbf{x})\rangle = \frac{1}{\rho(\mathbf{x})} \int_{H_{\mathbf{x}}} \left[\boldsymbol{\sigma}(\mathbf{x}') - \boldsymbol{\sigma}(\mathbf{x})\right] \rho(\mathbf{x}') \frac{\partial W(\mathbf{x} - \mathbf{x}')}{\partial \mathbf{x}} \, dv_{\mathbf{x}'} . \tag{6.26}$$

If the ratio of the densities is assumed to be one, the alternative formulation from Randles and Libersky (1996) is obtained, which is based on the rule given by Monaghan (1988)

$$\langle \operatorname{div} \boldsymbol{\sigma}(\mathbf{x}) \rangle = \int_{H_\mathbf{x}} \left[\boldsymbol{\sigma}(\mathbf{x}') - \boldsymbol{\sigma}(\mathbf{x}) \right] \frac{\partial W(\mathbf{x} - \mathbf{x}')}{\partial \mathbf{x}} \, dv_{\mathbf{x}'}. \tag{6.27}$$

With the representation of the derivative after (6.19)

$$\langle \operatorname{div} \boldsymbol{\sigma}(\mathbf{x}) \rangle = \frac{1}{\rho(\mathbf{x})} \int_{H_\mathbf{x}} W(\mathbf{x} - \mathbf{x}') \left[\operatorname{div} \left(\rho(\mathbf{x}') \boldsymbol{\sigma}(\mathbf{x}') \right) - \boldsymbol{\sigma}(\mathbf{x}) \operatorname{grad} \rho(\mathbf{x}') \right] dv_{\mathbf{x}'} \tag{6.28}$$

the condition from the integration constraint results to

$$\begin{aligned} &\int_{H_\mathbf{x}} W(\mathbf{x} - \mathbf{x}') \left[\operatorname{div} \left(\rho(\mathbf{x}') \boldsymbol{\sigma}(\mathbf{x}') \right) - \boldsymbol{\sigma}(\mathbf{x}) \operatorname{grad} \rho(\mathbf{x}') \right] dv_{\mathbf{x}'} \\ &= \int_{\partial H_\mathbf{x}} W(\mathbf{x} - \mathbf{x}') \rho(\mathbf{x}') \left[\boldsymbol{\sigma}(\mathbf{x}') - \boldsymbol{\sigma}(\mathbf{x}) \right] \mathbf{n}(\mathbf{x}') \, da_{\mathbf{x}'} \\ &- \int_{H_\mathbf{x}} \rho(\mathbf{x}') \left[\boldsymbol{\sigma}(\mathbf{x}') - \boldsymbol{\sigma}(\mathbf{x}) \right] \frac{\partial W(\mathbf{x} - \mathbf{x}')}{\partial \mathbf{x}'} \, dv_{\mathbf{x}'}. \end{aligned} \tag{6.29}$$

To meet the standard patch test, the above relation must be fulfilled for constant stresses. In this case, both sides are identical zero. This also applies to particles near or at the boundary and no resulting normal vector can be obtained.

6.4.2 Quotient Rule

Another way to rewrite the divergence of stress is based on the exploitation of the quotient rule. This approach goes back to Gingold and Monaghan (1982)

$$\operatorname{div} \left(\frac{1}{\rho} \boldsymbol{\sigma} \right) = \frac{1}{\rho} \operatorname{div} \boldsymbol{\sigma} - \frac{1}{\rho^2} \boldsymbol{\sigma} \operatorname{grad} \rho \quad \text{leading to} \quad \operatorname{div} \boldsymbol{\sigma} = \rho \operatorname{div} \left(\frac{1}{\rho} \boldsymbol{\sigma} \right) + \frac{1}{\rho} \boldsymbol{\sigma} \operatorname{grad} \rho. \tag{6.30}$$

In this case, the divergence is determined by the sum of stresses scaled by the density

$$\langle \operatorname{div} \boldsymbol{\sigma}(\mathbf{x}) \rangle = \rho(\mathbf{x}) \int_{H_\mathbf{x}} \left[\frac{\boldsymbol{\sigma}(\mathbf{x}')}{\rho(\mathbf{x}')^2} + \frac{\boldsymbol{\sigma}(\mathbf{x})}{\rho(\mathbf{x})^2} \right] \frac{\partial W(\mathbf{x} - \mathbf{x}')}{\partial \mathbf{x}} \rho(\mathbf{x}') \, dv_{\mathbf{x}'}. \tag{6.31}$$

A simplified form is obtained when the ratio of densities is again neglected, like in Bonet and Lok (1999)

$$\langle \operatorname{div} \boldsymbol{\sigma}(\mathbf{x}) \rangle = \int_{H_\mathbf{x}} \left[\boldsymbol{\sigma}(\mathbf{x}') + \boldsymbol{\sigma}(\mathbf{x}) \right] \frac{\partial W(\mathbf{x} - \mathbf{x}')}{\partial \mathbf{x}} \, dv_{\mathbf{x}'}. \tag{6.32}$$

Again, the relationship for the integration constraint must be adjusted. Based on (6.19) the divergence can be reformulated

$$\langle \operatorname{div} \boldsymbol{\sigma}(\mathbf{x}) \rangle = \rho(\mathbf{x}) \int_{H_{\mathbf{x}}} W(\mathbf{x} - \mathbf{x}') \left[\operatorname{div} \left(\frac{\boldsymbol{\sigma}(\mathbf{x}')}{\rho(\mathbf{x}')} \right) + \frac{\boldsymbol{\sigma}(\mathbf{x})}{\rho(\mathbf{x})^2} \operatorname{grad} \rho(\mathbf{x}') \right] dv_{\mathbf{x}'} \tag{6.33}$$

leading to the requirement

$$\begin{aligned} &\int_{H_{\mathbf{x}}} W(\mathbf{x} - \mathbf{x}') \left[\operatorname{div} \left(\frac{\boldsymbol{\sigma}(\mathbf{x}')}{\rho(\mathbf{x}')} \right) + \frac{\boldsymbol{\sigma}(\mathbf{x})}{\rho(\mathbf{x})^2} \operatorname{grad} \rho(\mathbf{x}') \right] dv_{\mathbf{x}'} \\ &= \int_{\partial H_{\mathbf{x}}} W(\mathbf{x} - \mathbf{x}') \left[\frac{\boldsymbol{\sigma}(\mathbf{x}')}{\rho(\mathbf{x}')} + \frac{\boldsymbol{\sigma}(\mathbf{x})}{\rho(\mathbf{x})^2} \rho(\mathbf{x}') \right] \mathbf{n}(\mathbf{x}') \, da_{\mathbf{x}'} \\ &- \int_{H_{\mathbf{x}}} \left[\frac{\boldsymbol{\sigma}(\mathbf{x}')}{\rho(\mathbf{x}')} + \frac{\boldsymbol{\sigma}(\mathbf{x})}{\rho(\mathbf{x})^2} \rho(\mathbf{x}') \right] \frac{\partial W(\mathbf{x} - \mathbf{x}')}{\partial \mathbf{x}'} \, dv_{\mathbf{x}'} . \end{aligned} \tag{6.34}$$

For the case of a constant stress and a constant density, the requirement reduces to the original variant given in (6.22)

$$\int_{H_{\mathbf{x}}} \frac{\partial W(\mathbf{x} - \mathbf{x}')}{\partial \mathbf{x}'} \, dv_{\mathbf{x}'} = \int_{\partial H_{\mathbf{x}}} W(\mathbf{x} - \mathbf{x}') \, \mathbf{n}(\mathbf{x}') \, da_{\mathbf{x}'} . \tag{6.35}$$

Remark 6.1 The second golden rule in SPH (Monaghan 1992) states that the formulation (6.27) should be used to calculate the derivative. This form guarantees that the derivative is zero if the corresponding quantities are constant in the neighborhood.

Remark 6.2 In Price (2012) a derivation of the mechanical balance equation for barotropic fluids from the discrete Euler-Lagrange equation is given. In Bonet and Lok (1999) the starting point is the discrete potential energy. The resulting discrete balance equations are equivalent to the formulation given in (6.31).

6.5 Reproducing Conditions

In SPH, W is usually a radial function. The reproducing conditions are not necessarily fulfilled in the discrete case, like e.g. for the 0th and the 1st order

$$\sum_{J=1}^{N_{\mathbf{x}}} W(\mathbf{x} - \mathbf{x}_J) \, v_J \neq 1, \qquad \sum_{J=1}^{N_{\mathbf{x}}} W(\mathbf{x} - \mathbf{x}_J) \, \mathbf{x}_J \, v_J \neq \mathbf{x}. \tag{6.36}$$

Due to the second property in (6.4), the weighting function satisfies the 0th order reproducing condition in the continuous case. However, in the discrete case this is mostly not given. The same follows for the derivatives

$$\sum_{J=1}^{N_\mathbf{x}} \frac{\partial W(\mathbf{x}-\mathbf{x}_J)}{\partial \mathbf{x}} v_J = -\sum_{J=1}^{N_\mathbf{x}} \frac{\partial W(\mathbf{x}_J-\mathbf{x})}{\partial \mathbf{x}_J} v_J \neq 0$$
$$\sum_{J=1}^{N_\mathbf{x}} \mathbf{x}_J \otimes \frac{\partial W(\mathbf{x}-\mathbf{x}_J)}{\partial \mathbf{x}} v_J = -\sum_{J=1}^{N_\mathbf{x}} \mathbf{x}_J \otimes \frac{\partial W(\mathbf{x}_J-\mathbf{x})}{\partial \mathbf{x}_J} v_J \neq \mathbf{1}. \tag{6.37}$$

Different approaches are available to restore the requirements. In the following, four different variants are presented. The first two allow a complete fulfillment of the reproducing conditions. The last two forms show frequently used corrections for certain requirements. In addition, other approaches exist. An older overview can be found in Belytschko et al. (1998).

Variant 1 Since the shape functions are based on an integral operator, the approaches from Sect. 4.6.4 can be directly applied to satisfy the reproducing conditions up to the necessary order. The component ω assuming $\rho = 1$ in (4.140) can be identified as the weighting function W. The remaining factor corresponds to the correction factor

$$\tilde{W}(\mathbf{x}-\mathbf{x}_J) = C(\mathbf{x}, \mathbf{x}_J)\, W(\mathbf{x}-\mathbf{x}_J), \quad C(\mathbf{x}, \mathbf{x}_J) = \mathbf{p}(0) \cdot \mathbf{M}(\mathbf{x})\, \widetilde{\mathbf{p}}(\mathbf{x}-\mathbf{x}_J). \tag{6.38}$$

The weighting function that fulfills the reproducing conditions is marked with a tilde over W.

Variant 2 An alternative approach is provided by Bonet and Lok (1999). The whole algorithm is indicated as Corrected Smoothed Particle Hydrodynamics (CSPH). It is based on the previous variant, considering only reproducibility up to 1st order. This allows the correction scheme to be simplified

$$\tilde{W}(\mathbf{x}-\mathbf{x}_J) = W(\mathbf{x}-\mathbf{x}_J)\, \alpha(\mathbf{x}) \left[1 + \boldsymbol{\beta}(\mathbf{x}) \cdot (\mathbf{x}-\mathbf{x}_J)\right]. \tag{6.39}$$

The values of the vector $\boldsymbol{\beta}$ are determined by the restoration of the 1st order reproducing condition

$$\boldsymbol{\beta}(\mathbf{x}) = \left[\sum_{J=1}^{N_\mathbf{x}} [\mathbf{x}-\mathbf{x}_J] \otimes [\mathbf{x}-\mathbf{x}_J]\, W(\mathbf{x}-\mathbf{x}_J)\, v_J\right]^{-1} \sum_{J=1}^{N_\mathbf{x}} W(\mathbf{x}-\mathbf{x}_J)\, [\mathbf{x}_J - \mathbf{x}]\, v_J. \tag{6.40}$$

After the calculation of $\boldsymbol{\beta}$, the parameter α is determined from the fulfillment of the 0th order reproducing condition

$$\alpha(\mathbf{x}) = \frac{1}{\sum_{J=1}^{N_\mathbf{x}} \left[1 + \boldsymbol{\beta}(\mathbf{x}) \cdot (\mathbf{x}-\mathbf{x}_J)\right] W(\mathbf{x}-\mathbf{x}_J)\, v_J}. \tag{6.41}$$

Variant 3 If only the 1st order reproducing condition in the derivatives has to be satisfied, the alternative correction from Randles and Libersky (1996) can be used, see also Bonet and Lok (1999). A normalization guarantees that the reproducing condition of 0th order is satisfied

$$\tilde{W}(\mathbf{x}-\mathbf{x}_J)=\frac{W(\mathbf{x}-\mathbf{x}_J)}{\sum_{I=1}^{N_\mathbf{x}} W(\mathbf{x}-\mathbf{x}_I)\,v_I}. \tag{6.42}$$

The requirement on the 1st order in the derivative can be recovered using the tensor **L** analogous to Sect. 4.6.2

$$\frac{\partial \tilde{W}^c(\mathbf{x}-\mathbf{x}_J)}{\partial \mathbf{x}}=\frac{\partial \tilde{W}(\mathbf{x}-\mathbf{x}_J)}{\partial \mathbf{x}}\mathbf{L}^{-1}(\mathbf{x}). \tag{6.43}$$

Since the weighting function $\tilde{W}$ satisfies the reproducing condition of 0th order in the derivatives, the tensor **L** can be alternatively formulated based on differences

$$\mathbf{L}(\mathbf{x})=\sum_{J=1}^{N_\mathbf{x}} \mathbf{x}_J\otimes\frac{\partial \tilde{W}(\mathbf{x}-\mathbf{x}_J)}{\partial \mathbf{x}}v_J=\sum_{J=1}^{N_\mathbf{x}}(\mathbf{x}_J-\mathbf{x})\otimes\frac{\partial \tilde{W}(\mathbf{x}-\mathbf{x}_J)}{\partial \mathbf{x}}v_J. \tag{6.44}$$

Variant 4 In Monaghan (1988), the 0th order reproducing condition in the derivatives is restored by a symmetrization. If a constant displacement field is present, by definition, the gradient is always equal to zero

$$\operatorname{grad}\mathbf{u}(\mathbf{x})=\sum_{J=1}^{N_\mathbf{x}}\left[\mathbf{u}_J-\mathbf{u}(\mathbf{x})\right]\frac{\partial W(\mathbf{x}-\mathbf{x}_J)}{\partial \mathbf{x}}v_J. \tag{6.45}$$

This relationship corresponds to the product rule given in (6.27).

6.6 Discrete Conservation Properties

In the SPH, the discretization of the divergence operator is often based on either the product or the quotient rule. In these cases, the conditions differ from the requirements in Sect. 4.5.5. Therefore, a separate investigation on the conservation of global balance equations is needed. For the mechanical case, the study is based on the internal force due to the state of stress inside the material. The derivation focuses on the strong form with reduced order but can be applied to the other descriptions in the same manner.

6.6.1 *Product Rule*

Based on (6.26), the internal force at the particle is determined from a difference

$$\mathbf{f}_I^s = -v_I \sum_{J=1}^{N_I} \frac{\rho_J}{\rho_I} \left[\boldsymbol{\sigma}_J - \boldsymbol{\sigma}_I\right] \frac{\partial W\left(\mathbf{x}_I - \mathbf{x}_J\right)}{\partial \mathbf{x}_I} v_J. \tag{6.46}$$

Exploiting (6.12), this contribution can also be understood as the sum of forces between individual particles

$$\mathbf{f}_I^s = -\sum_{J=1}^{N_I} \frac{\rho_J}{\rho_I} \left[\mathbf{f}_{IJ}^p + \mathbf{f}_{JI}^p\right], \quad \mathbf{f}_{IJ}^p = \boldsymbol{\sigma}_J \frac{\partial W\left(\mathbf{x}_I - \mathbf{x}_J\right)}{\partial \mathbf{x}_I} v_I v_J. \tag{6.47}$$

The alternative formulation given in Randles and Libersky (1996) results if the ratio between the densities is assumed to be one.

Global balance of momentum. This requirement is satisfied if the internal forces over all particles n sum to zero, see also Sect. 4.5.5. By substituting (6.47) into (4.86), since $W\left(\mathbf{x} - \mathbf{x}_J\right) \equiv 0$, if $\mathbf{x} \notin H_{\mathbf{x}}$, and by exploiting the interchange of indices, the condition can be specified in more detail

$$\sum_{I=1}^{n} \mathbf{f}_I^s = -\sum_{I=1}^{n} \sum_{J=1}^{n} \left[\frac{\rho_J}{\rho_I} + \frac{\rho_I}{\rho_J}\right] \mathbf{f}_{IJ}^p = \sum_{J=1}^{n} v_J \boldsymbol{\sigma}_J \sum_{I=1}^{N_J} \left[\frac{\rho_J}{\rho_I} + \frac{\rho_I}{\rho_J}\right] \frac{\partial W\left(\mathbf{x}_I - \mathbf{x}_J\right)}{\partial \mathbf{x}_I} v_I = 0. \tag{6.48}$$

To satisfy conservation of momentum for arbitrary stresses, the inner loop must be equal to zero

$$\sum_{I=1}^{N_J} \left[\frac{\rho_J}{\rho_I} + \frac{\rho_I}{\rho_J}\right] \frac{\partial W\left(\mathbf{x}_I - \mathbf{x}_J\right)}{\partial \mathbf{x}_I} v_I = 0. \tag{6.49}$$

In the version of Randles and Libersky (1996), the condition reduces to satisfying the 0th order reproducing condition in the derivatives.

Global balance of angular momentum. This conservation results in the requirement that the sum over all moments due to the internal forces must be equal to zero

$$\sum_{I=1}^{n} \mathbf{f}_I^s \times \mathbf{x}_I = \sum_{I=1}^{n} \mathcal{E} : \mathbf{x}_I \otimes \mathbf{f}_I^s = 0. \tag{6.50}$$

Using (6.47) and taking advantage of the interchangeability of indices, the condition can be formulated in dependence on the addition of position vectors

$$\begin{aligned} &\sum_{I=1}^{n} \sum_{J=1}^{n} \mathcal{E} : \mathbf{x}_I \otimes \frac{\rho_J}{\rho_I} \left[\mathbf{f}_{IJ}^p + \mathbf{f}_{JI}^p\right] \\ &= \sum_{I=1}^{n} \sum_{J=1}^{n} \mathcal{E} : \left[\frac{\rho_J}{\rho_I} \mathbf{x}_I + \frac{\rho_I}{\rho_J} \mathbf{x}_J\right] \otimes \frac{\partial W\left(\mathbf{x}_I - \mathbf{x}_J\right)}{\partial \mathbf{x}_I} \boldsymbol{\sigma}_J v_I v_J = 0. \end{aligned} \tag{6.51}$$

It has to be noted that the symmetry of the stress tensor is assumed in the derivation, i.e. $\mathcal{E} : \boldsymbol{\sigma} \equiv 0$. In order to preserve angular momentum, the following condition must be fulfilled

$$\sum_{I=1}^{N_J} \left[\frac{\rho_J}{\rho_I} \mathbf{x}_I + \frac{\rho_I}{\rho_J} \mathbf{x}_J \right] \otimes \frac{\partial W(\mathbf{x}_I - \mathbf{x}_J)}{\partial \mathbf{x}_I} v_I = \mathbf{1}. \tag{6.52}$$

If the ratio of the densities is again set to one, the requirement reduces to satisfying the 0th and 1st order reproducing conditions in the derivatives.

6.6.2 *Quotient Rule*

Using the derivative rule (6.31) another variant for calculating the internal force at a particle is obtained

$$\mathbf{f}_I^s = -v_I \rho_I \sum_{J=1}^{N_I} \left[\frac{\boldsymbol{\sigma}_J}{\rho_J^2} + \frac{\boldsymbol{\sigma}_I}{\rho_I^2} \right] \frac{\partial W(\mathbf{x}_I - \mathbf{x}_J)}{\partial \mathbf{x}_I} \rho_J v_J. \tag{6.53}$$

Taking advantage of (6.12), the force at a particle can also be interpreted as a result of the difference of forces between particles

$$\mathbf{f}_I^s = -\sum_{J=1}^{N_I} [\mathbf{f}_{IJ} - \mathbf{f}_{JI}], \quad \mathbf{f}_{IJ} = \rho_I \rho_J \frac{\boldsymbol{\sigma}_I}{\rho_I^2} \frac{\partial W(\mathbf{x}_I - \mathbf{x}_J)}{\partial \mathbf{x}_I} v_I v_J. \tag{6.54}$$

The contribution $\mathbf{f}_{IJ}$ corresponds to the force acting from particle J on particle I, whereas $\mathbf{f}_{JI}$ describes the inverse relation. Another variation results when the ratio of the densities is again set to one

$$\mathbf{f}_I^s = -\sum_{J=1}^{N_I} \left[\boldsymbol{\sigma}_I \frac{\partial W(\mathbf{x}_I - \mathbf{x}_J)}{\partial \mathbf{x}_I} v_J v_I - \boldsymbol{\sigma}_J \frac{\partial W(\mathbf{x}_J - \mathbf{x}_I)}{\partial \mathbf{x}_J} v_I v_J \right] = -\sum_{J=1}^{N_I} [\mathbf{f}_{IJ} - \mathbf{f}_{JI}]. \tag{6.55}$$

The only difference between both variants is the calculation of $\mathbf{f}_{IJ}$. Here it has to be noted that (6.55) using (6.6) corresponds exactly to (4.34), which is derived from the weak form based on differences.

Global balance of momentum. If particle J is in the neighborhood of the particle I and vice versa I is in the neighborhood of J, the global conservation of momentum is always satisfied. Since $W(\mathbf{x} - \mathbf{x}_J) \equiv 0$, if $\mathbf{x} \notin H_{\mathbf{x}}$, the sums can be formed over all particles and the indices can be interchanged

$$\sum_{I=1}^{n} \mathbf{f}_I^s = -\sum_{I=1}^{n} \sum_{J=1}^{n} [\mathbf{f}_{IJ} - \mathbf{f}_{JI}] = -\sum_{I=1}^{n} \sum_{J=1}^{n} [\mathbf{f}_{IJ} - \mathbf{f}_{IJ}] \equiv 0. \tag{6.56}$$

Global balance of angular momentum. Since the indices can be arbitrarily exchanged, this proof can also be formulated in terms of a difference of position vectors. Substituting (6.54) into the general condition (4.89) leads to the alternative requirement

$$\sum_{I=1}^{n} \mathcal{E} : \mathbf{x}_I \otimes \mathbf{f}_I^s = -\sum_{I=1}^{n}\sum_{J=1}^{n} \mathcal{E} : \mathbf{x}_I \otimes (\mathbf{f}_{IJ} - \mathbf{f}_{JI}) = -\sum_{I=1}^{n}\sum_{J=1}^{n} \mathcal{E} : (\mathbf{x}_I - \mathbf{x}_J) \otimes \mathbf{f}_{IJ} = 0. \tag{6.57}$$

Unlike the conservation of momentum, this condition is not satisfied per se. By substituting $\mathbf{f}_{IJ}$ from (6.54) into (6.57) and after some algebra, a requirement can be stated

$$-\sum_{I=1}^{n} \mathcal{E} : \left(\sum_{J=1}^{N_I} (\mathbf{x}_I - \mathbf{x}_J) \otimes \frac{\partial W(\mathbf{x}_I - \mathbf{x}_J)}{\partial \mathbf{x}_I} \rho_J v_J \right) \frac{v_I}{\rho_I} \boldsymbol{\sigma}_I = 0. \tag{6.58}$$

Again, the symmetry of the stress tensor $\boldsymbol{\sigma}$ is exploited. The global angular momentum is conserved only if the contribution in the parenthesis leads to the unit tensor

$$\sum_{J=1}^{N_I} (\mathbf{x}_I - \mathbf{x}_J) \otimes \frac{\partial W(\mathbf{x}_I - \mathbf{x}_J)}{\partial \mathbf{x}_I} \rho_J v_J = \mathbf{1}. \tag{6.59}$$

If the ratio of densities is set to one, the requirement results in the fulfillment of the 0th and 1st order reproducing conditions in the derivatives.

Remark. If the stress tensor is isotropic, i.e., no shear stresses occur, then, in addition to the global conservation of momentum, the condition on angular momentum is always satisfied. In this case, (6.57) changes to

$$\sum_{I=1}^{n}\sum_{J=1}^{n} \mathcal{E} : (\mathbf{x}_I - \mathbf{x}_J) \otimes \mathbf{f}_{IJ} = \sum_{I=1}^{n}\sum_{J=1}^{n} v_I v_J \rho_J \frac{p_I}{\rho_I} \frac{\partial W(\mathbf{x}_I - \mathbf{x}_J)}{\partial \mathbf{x}_I} \times (\mathbf{x}_I - \mathbf{x}_J) = 0. \tag{6.60}$$

Since W is based on a radial function

$$\frac{\partial W(\mathbf{x}_I - \mathbf{x}_J)}{\partial \mathbf{x}_I} = \frac{\partial W}{\partial \|\mathbf{x}_I - \mathbf{x}_J\|} \frac{\mathbf{x}_I - \mathbf{x}_J}{\|\mathbf{x}_I - \mathbf{x}_J\|}, \tag{6.61}$$

the force is always parallel to the distance between the particles. The cross product between two vectors pointing in the same direction is by definition equal to zero and angular momentum is always conserved

$$\sum_{I=1}^{n}\sum_{J=1}^{n} v_I v_J \rho_J \frac{p_I}{\rho_I} \frac{\partial W}{\partial \|\mathbf{x}_I - \mathbf{x}_J\|} \left(\frac{\mathbf{x}_I - \mathbf{x}_J}{\|\mathbf{x}_I - \mathbf{x}_J\|} \times (\mathbf{x}_I - \mathbf{x}_J) \right) \equiv 0. \tag{6.62}$$

6.7 Search Algorithm

The first step in the SPH is the determination of the neighborhoods. Usually, an algorithm seeks the nearest particles. If the distance of particle J to particle I is smaller than a given value, then J is in the neighborhood of I. Normally, the radius correlates with the smoothing length (Li and Liu 2007). Mostly, r is in the range of 1.3–2.0 times the initial particle distance (Fig. 6.1). Since the body is usually divided into grid-like individual particles, the radius is constant for all particles.

A redefinition of the neighborhood during the simulation can lead to a change in the quantities, although there is no external load. In order to avoid this unphysical behavior, the smoothing length can be adjusted. Monaghan (2002) suggests that the density must be the same before and after the redetermination

$$\rho_I = \sum_{J=1}^{N_I} m_J \, W\left(\mathbf{x}_I - \mathbf{x}_J, h_I\right), \quad \text{with} \quad h_I := \left(\frac{m_I}{\rho_I}\right)^{\frac{1}{3}} = (v_I)^{\frac{1}{3}}. \tag{6.63}$$

Since the density is a function of the smoothing length, a nonlinear system of equations results. For its solution the Newton-Raphson iteration can be applied. The dependence of the weighting function on h must also be considered in the calculation of the force in order not to violate the balance of momentum (Nelson and Papaloizou 1994). For a varying smoothing length an additional term Ω appears in the calculation of the forces, like for the approach from (6.55)

$$\mathbf{f}_I^s = -\sum_{J=1}^{N_I} \left[\frac{\boldsymbol{\sigma}_I}{\Omega_I} \frac{\partial W(\mathbf{x}_I - \mathbf{x}_J, h_I)}{\partial \mathbf{x}_I} - \frac{\boldsymbol{\sigma}_J}{\Omega_J} \frac{\partial W(\mathbf{x}_J - \mathbf{x}_I, h_J)}{\partial \mathbf{x}_J}\right] v_I v_J. \tag{6.64}$$

The additional quantity Ω contains the derivative of the weighting function with respect to the smoothing length

$$\Omega_I = 1 + \frac{h_I}{3\,\rho_I} \sum_{J=1}^{N_I} m_J \frac{\partial W(\mathbf{x}_I - \mathbf{x}_J, h_I)}{\partial h_I}. \tag{6.65}$$

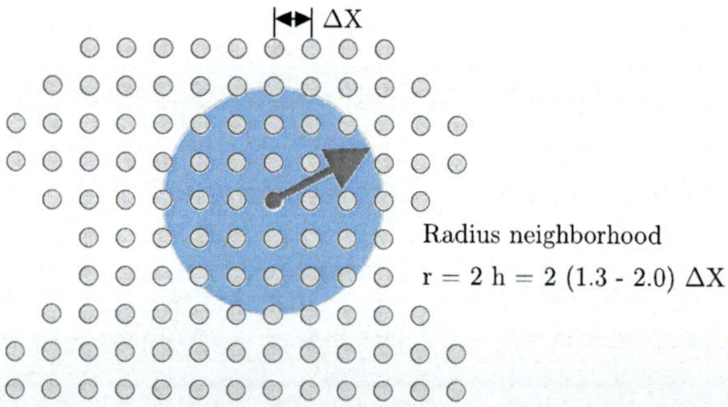

Fig. 6.1 Initial subdivision of the domain into a grid like distribution of particles and determination of the neighborhood by means of a search radius

A detailed derivation of this equation can be found in Price and Monaghan (2004) or Price (2012).

Another condition on the search algorithm follows from (6.56). If particle J lies in the neighborhood of particle I and particle I is also part of the neighborhood of J, global momentum is conserved. If the search radii are the same for each particle, this requirement is always met, see also Sect. 7.5.

In addition, the particle distribution in the neighborhood should be statistically homogeneous. When simulating fluids, the neighborhood of the particles is redetermined at every time step. Disturbed particle distributions can occur, resulting in penetration and a mixture of particles. To prevent this, an alternative approach is used in Monaghan (1989). The position vector of the particle $\mathbf{x}_I$ is not updated by the solution function $\mathbf{v}_I$ of the differential equation, but via a smoothing velocity

$$\dot{\mathbf{x}}_I = \mathbf{v}_I + \xi \sum_{J=1}^{N_I} \frac{m_J}{\rho_{IJ}} (\mathbf{v}_J - \mathbf{v}_I)\, W(\mathbf{x}_I - \mathbf{x}_J), \quad \rho_{IJ} = \frac{1}{2} (\rho_I + \rho_J). \tag{6.66}$$

The current coordinate is calculated from the integration of $\dot{\mathbf{x}}_I$. The parameter ξ is usually selected between 0 and 0.5. This approach does not affect the balance of momentum. If this correction is applied the algorithm is also called XSPH.

Alternatively, there are other correction schemes that allow an improved particle distribution. These are based either on remeshing techniques (Xu et al. 2009), the introduction of an advection velocity (Adami et al. 2013), or on an Arbitrary Lagrangian Eulerian (ALE) approach (Oger et al. 2016).

Remark. The reasons for the disturbed particle distribution are not investigated in the presented algorithms. Instead, only attempts are made to correct the symptom. As shown in Sect. 6.13, rank instability is a central topic in SPH. In addition, configurational consistency plays a role, which has not yet been investigated in the SPH. Only if all negative effects due to a violation of the requirements listed in Sect. 4.5 can be excluded, investigations on the causes of disturbed particle distributions are possible.

6.8 Tensile Instability

The SPH in its original form is based on the deformed configuration. As shown in Swegle et al. (1995), to avoid a tensile instability it must be ensured that the product of the second derivative of the weighting function with the stress at any particle J is always less than zero. In Monaghan (2000), an approach based on artificial stresses is introduced to avoid negative effects due to tensile instability.

6.9 SPH for Solids

In many applications, the modeling of solids by means of SPH is based on the rule (6.55), see e.g. Ganzenmüller (2015) or Rabczuk et al. (2004). In order to avoid tensile instabilities, the equations are discretized with respect to the initial configuration. Mostly, the 1st Piola Kirchhoff stress tensor $\mathbf{P}$ is used instead of $\boldsymbol{\sigma}$. Hence, the current volume v must be replaced by the initial volume V and the weighting function must be evaluated with respect to $\mathbf{X}$

$$\mathbf{f}_I^s = -\sum_{J=1}^{N_I} \left[\mathbf{P}_I \frac{\partial W\left(\mathbf{X}_I - \mathbf{X}_J\right)}{\partial \mathbf{X}_I} - \mathbf{P}_J \frac{\partial W\left(\mathbf{X}_J - \mathbf{X}_I\right)}{\partial \mathbf{X}_J} \right] V_I V_J. \tag{6.67}$$

For hyperelastic materials, $\mathbf{P}$ is calculated from the derivative of the specific strain energy function with respect to the deformation gradient

$$\mathbf{P}_I = \rho_0 \frac{\partial \psi(\mathbf{F}_I)}{\partial \mathbf{F}_I}. \tag{6.68}$$

To ensure that a constant displacement does not lead to strains in the body, the rule (6.45) is mostly used to compute $\mathbf{F}$. In addition, the derivative of the weighting function is corrected based on (6.44)

$$\begin{aligned} \mathbf{F}_I &= \mathbf{1} + \sum_{J=1}^{N_I} \left(\mathbf{u}_J - \mathbf{u}_I\right) \otimes \frac{\partial W(\mathbf{X}_I - \mathbf{X}_J)}{\partial \mathbf{X}_I} \mathbf{L}_I{}^{-1} V_J, \\ \mathbf{L}_I &= \sum_{J=1}^{N_I} \left(\mathbf{X}_J - \mathbf{X}_I\right) \otimes \frac{\partial \tilde{W}\left(\mathbf{X}_I - \mathbf{X}_J\right)}{\partial \mathbf{X}_I} V_J. \end{aligned} \tag{6.69}$$

Hence, the reproducing condition of 1st order in the derivatives is satisfied. The corrected derivative is also mostly applied to calculate the internal forces

$$\mathbf{f}_I^s = -\sum_{J=1}^{N_I} \left[\mathbf{P}_I \frac{\partial W\left(\mathbf{X}_I - \mathbf{X}_J\right)}{\partial \mathbf{X}_I} \mathbf{L}_I{}^{-1} - \mathbf{P}_J \frac{\partial W\left(\mathbf{X}_J - \mathbf{X}_I\right)}{\partial \mathbf{X}_J} \mathbf{L}_J{}^{-1} \right] V_I V_J. \tag{6.70}$$

Due to the corrected weighting function, and due to the formulation based on the quotient rule, momentum but not angular momentum is conserved.

6.10 SPH for Fluids

In the calculation of fluid flows, the neighborhood is always redefined in order to simulate very large deformations. The equations are formulated with respect to the deformed configuration. Since the molten metal can be regarded as incompressible, only volume preserving fluids are considered. The discretization is based on the differential equation (3.14). For the pressure gradient the rule (6.55) is often used (Colagrossi and Landrini 2003; Fürstenau et al. 2019)

$$\mathbf{f}_I^s = \mathbf{f}_I^p + \mathbf{f}_I^v, \qquad \mathbf{f}_I^p = \sum_{J=1}^{N_I} [p_J + p_I] \frac{\partial W(\mathbf{x}_I - \mathbf{x}_J)}{\partial \mathbf{x}_I} v_I v_J. \tag{6.71}$$

Since no shear stresses occur, the global balance of momentum and angular momentum are always fulfilled. The pressure can be calculated in two different ways. The first approach is based on the direct fulfillment of the incompressibility condition. The direct solution of this constraint leads to a clustering of particles. Alternatively, in the SPH p can be solved by the Pressure Poisson equation (Cummins and Rudman 1999; Shao and Lo 2003 or Hu and Adams 2007). The particle velocities and position vectors are calculated subsequently. This scheme is called Incompressible Smoothed Particle Hydrodynamics (ISPH) and describes an own area within the SPH that is not considered here. The second option describes the material as a barotropic fluid. The pressure calculates from the density like presented in MacDonald (1966)

$$p_I = \frac{\rho_I\, c_0^2}{7} \left[\left(\frac{\rho_I}{\rho_0} \right)^7 - 1 \right]. \tag{6.72}$$

The velocity of sound c_0 in the liquid can be interpreted as a penalty parameter that forces the density to be equal to its initial value. For this reason, the alternative approach $p_I = c_0^2(\rho_I - \rho_0)$ can also be used. The penalty parameter should be 10 times larger than the maximum velocity component ($c_0 \geq 10 \max v_i$) to keep the density changes in the area below one percent (Monaghan 2005). Since the incompressibility constraint is not enforced exactly, this approach is called weakly compressible. The density of a particle is calculated from the discretized continuity equation. Often the rule (6.45) is used, which ensures a constant density, if the velocity is constant

$$\dot{\rho}_I = -\rho_I \sum_{J=1}^{N_I} (\mathbf{v}_J - \mathbf{v}_I) \cdot \frac{\partial W(\mathbf{x}_I - \mathbf{x}_J)}{\partial \mathbf{x}_I} v_J. \tag{6.73}$$

In the SPH for fluids, the strong form is discretized directly. The required second derivative of the weighting function in the viscous term can lead to unstable solutions. Therefore, an approach based on two first derivatives is mostly applied

$$\mathbf{f}_I^v = -\eta\, v_I \sum_{J=1}^{N_I} \frac{\rho_I + \rho_J}{\rho_I} \frac{(\mathbf{v}_I - \mathbf{v}_J) \cdot (\mathbf{x}_I - \mathbf{x}_J)}{\|\mathbf{x}_I - \mathbf{x}_J\|^2 + \epsilon^2} \frac{\partial W(\mathbf{x}_I - \mathbf{x}_J)}{\partial \mathbf{x}_I} v_J. \tag{6.74}$$

This form goes back to Brookshaw (1985), and was extended in Espanol and Revenga (2003) to general tensors. By a small modification it can be achieved that this approach conserves global momentum as well as angular momentum. If the density in the denominator is assumed to be constant, the equation can be formulated as a function of the kinematic viscosity ν, i.e. $\eta = \nu\,\rho$. Taking advantage of (6.12), $\mathbf{f}_I^\upsilon$ can again be expressed as the difference of forces between individual particles

$$\mathbf{f}_I^\upsilon = -\sum_{J=1}^{N_I} \left[\mathbf{f}_{IJ}^\upsilon - \mathbf{f}_{JI}^\upsilon\right], \quad \mathbf{f}_{IJ}^\upsilon = \nu \upsilon_I \rho_I \frac{(\mathbf{v}_I - \mathbf{v}_J) \cdot (\mathbf{x}_I - \mathbf{x}_J)}{\|\mathbf{x}_I - \mathbf{x}_J\|^2 + \epsilon^2} \frac{\partial W\,(\mathbf{x}_I - \mathbf{x}_J)}{\partial \mathbf{x}_I} \upsilon_J. \tag{6.75}$$

Analogous to (6.56), this ensures that global momentum is conserved. Since the quantities are scalars, statement (6.62) is valid and the condition on the global conservation of angular momentum is also fulfilled.

6.11 Boundary Conditions

The neighborhood of particles at the boundary is not completely filled. Most shape functions in SPH neither fulfill the reproducing conditions nor possess the Kronecker-δ property. If a constant quantity is approximated using the standard approach, differences can be observed between the values at inner and outer particles (Takeda et al. 1994). In the SPH, a lot of methods for imposing boundary conditions have been established, which differ strongly from standard Galerkin schemes.

For the imposition of Dirichlet boundary conditions, four different approaches are available. Two of them are based on the introduction of so-called ghost particles, which lie outside the area (Libersky et al. 1993) and represent a fixed border (Fig. 6.2). The particles are not taken into account in the calculation but lead to a completely filled neighborhood for particles at or near the boundary. This form is often used in the simulation of solids (Ganzenmüller 2015). Randles and Libersky (1996) presented a variant in which the ghost particles are canceled out. For solids, the Dirichlet boundary condition is assigned to all ghost particles. The number and distribution of these particles result from the condition of a filled neighborhood. In fluid-structure

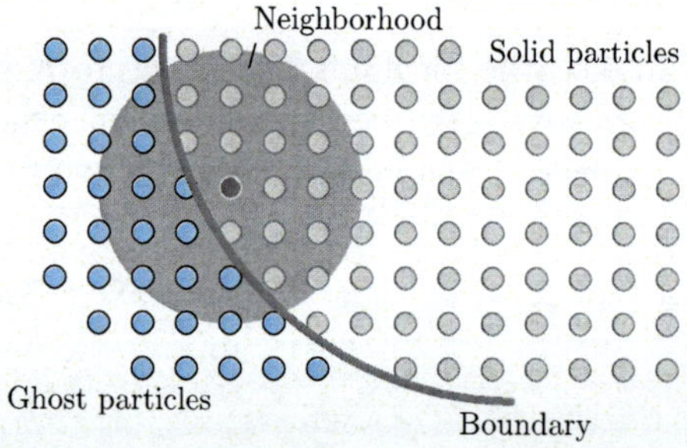

Fig. 6.2 Introduction of ghost particles to impose boundary conditions

interactions with rigid walls, similar approaches for the imposition of boundary conditions are available (Morris et al. 1997; Adami et al. 2012). To fulfill the normal and tangential boundary conditions for fluids from (3.16) and (3.17) suitable values must be defined at the ghost particles.

A second, similar option is the mirroring of particles along the boundary, as suggested in Colagrossi and Landrini (2003) or Macia et al. (2011). This also leads to ghost particles outside the area. Suitable values must be defined at these particles, to impose the boundary conditions accurately.

Alternatively, the Dirichlet boundary conditions can be replaced by forces. Estimates can be based on the Lennard Jones potential (Monaghan 1994) or on radial functions (Monaghan and Kajtar 2009). In Bonet and Kulasegaram (2000) the penalty method is applied to impose Dirichlet boundary conditions. The prerequisite for this approach is the fulfillment of the reproducing conditions at the boundary.

A fourth possibility is the direct integration of the boundary conditions into the differential equations. Corresponding approaches can be found in Kulasegaram et al. (2004), Macia et al. (2012) or Ferrand et al. (2013).

Neumann boundary conditions are usually imposed as particle forces. Since no direct information about the surface is available in SPH, individual surface portions must be assigned to each particle. This area can be estimated e.g. from (6.63) as suggested in Soleimani (2017)

$$a_I = h_I^2 = (v_I)^{\frac{2}{3}} . \tag{6.76}$$

Alternatively, the area belonging to a particle can be determined using the weighting function (Antoci et al. 2007). The force is introduced into the body via a layer of particles at the boundary. Another approach for imposing Dirichlet and Neumann boundary conditions based on fixed ghost particles can be found in Marrone et al. (2011).

A study that compares different approaches for fluid-structure interactions can be found in Valizadeh and Monaghan (2015). This investigation concludes that the modeling based on fixed ghost particles leads to the most physically meaningful results.

6.12 Oscillations

The simulation of shock waves or convection-dominated flows can show artificial oscillations in the velocity and pressure field. Therefore an artificial viscosity is introduced in Monaghan and Gingold (1983) to dampen this oscillation

$$\mathbf{f}_I^{art\,vis} = -v_I \alpha\, h\, c_0 \sum_{J=1}^{N_I} \frac{\rho_I \rho_J}{2(\rho_I + \rho_J)} \frac{(\mathbf{v}_I - \mathbf{v}_J) \cdot (\mathbf{x}_I - \mathbf{x}_J)}{\|\mathbf{x}_I - \mathbf{x}_J\|^2 + \epsilon^2} \frac{\partial W(\mathbf{x}_I - \mathbf{x}_J)}{\partial \mathbf{x}_I} v_J . \tag{6.77}$$

In Molteni and Colagrossi (2009) it was observed that despite the artificial viscosity high-frequency disturbances still occur in the pressure field. By the introduction of an additional diffusion term in the continuity equation, these oscillations could be damped out (Molteni and Colagrossi 2009; Marrone et al. 2011)

$$\rho_I^{art\,dens} = \delta\, h\, \bar{c}_0 \sum_{J=1}^{N_I} \left[2\,(\rho_J - \rho_I)\,\frac{\mathbf{x}_J - \mathbf{x}_I}{\|\mathbf{x}_I - \mathbf{x}_J\|^2} - \operatorname{grad}\rho_I - \operatorname{grad}\rho_J \right] \cdot \frac{\partial W\,(\mathbf{x}_I - \mathbf{x}_J)}{\partial \mathbf{x}_I} v_J . \tag{6.78}$$

In the discretization of the gradient, the corrected derivative using the tensor **L** (6.43) is used

$$\operatorname{grad}\rho_I = \sum_{J=1}^{N_I} (\rho_J - \rho_I)\,\frac{\partial W_I\,(\mathbf{x}_I - \mathbf{x}_J)}{\partial \mathbf{x}_I} \mathbf{L}_I{}^{-1} . \tag{6.79}$$

The application of this stabilization is known as δ-SPH. The numerical parameters α and δ control the strength of the damping and are selected much smaller than 1.

An alternative approach is based on a formulation of the equations in a moving frame of reference. This scheme corresponds to a coupling of the SPH with the Arbitrary Lagrangian-Eulerian formulation (Vila 1999). An application can be found in Marongiu et al. (2010). Since this method requires the solution of the non-linear Riemann problem, this variant is also called Riemann SPH (Oger et al. 2016). A comparison of different approaches can be found in Antuono et al. (2012).

6.13 Rank Instability

Belytschko et al. (2000) shows that artificial zero-energy modes also occur in the SPH. The reasons lie in the reduced number of evaluation points, which is limited to the number of particles. By introducing additional integration points, which are called stress points in SPH, the negative effects due to under-integration can be diminished (Vignjevic et al. 2000; Belytschko et al. 2000).

The approach of using stress points in SPH goes back to Dyka and Ingel (1995) and Dyka et al. (1997) and was initially developed to correct the tensile instability.

To avoid these additional evaluation points, an alternative approach is proposed in Bonet and Kulasegaram (2000). This correction is based on a gradient type stabilization similar to the concept introduced in Sect. 5.1.2.

A third option is suggested in Ganzenmüller (2015) which is already discussed in Sect. 5.2.4. Thereby, an additional force is introduced into the system, which actually corrects the violation of the configurational consistency from (4.84) by employing the penalty method

$$\mathbf{f}_I^{cor} = -V_I \sum_{J=1}^{N_I} \alpha \frac{E}{\|\mathbf{X}_I - \mathbf{X}_J\|^2} \epsilon_{IJ} \frac{\mathbf{x}_I - \mathbf{x}_J}{\|\mathbf{x}_I - \mathbf{x}_J\|} W\,(\mathbf{X}_I - \mathbf{X}_J)\, V_J . \tag{6.80}$$

The error from the configurational consistency ϵ_{IJ} is determined separately for each particle pair

$$\epsilon_{IJ} = \left[(\mathbf{x}_I - \mathbf{x}_J) - \frac{1}{2} (\mathbf{F}_I + \mathbf{F}_J) (\mathbf{X}_I - \mathbf{X}_J) \right] \cdot \frac{\mathbf{x}_I - \mathbf{x}_J}{\|\mathbf{x}_I - \mathbf{x}_J\|}. \quad (6.81)$$

This approach is developed for simulating solid deformations and is formulated with respect to the initial configuration. The size α corresponds to the penalty parameter and is set in relation to Young's modulus E (Ganzenmüller 2015).

References

S. Adami, X.Y. Hu, N.A. Adams, A generalized wall boundary condition for smoothed particle hydrodynamics. J. Comput. Phys. **231**(21), 7057–7075 (2012)

S. Adami, X.Y. Hu, N.A. Adams, A transport-velocity formulation for smoothed particle hydrodynamics. J. Comput. Phys. **241**, 292–307 (2013)

C. Antoci, M. Gallati, S. Sibilla, Numerical simulation of fluid-structure interaction by SPH. Comput. Struct. **85**(11–14), 879–890 (2007)

M. Antuono, A. Colagrossi, S. Marrone, Numerical diffusive terms in weakly-compressible SPH schemes. Comput. Phys. Commun. **183**(12), 2570–2580 (2012)

T. Belytschko, Y. Krongauz, J. Dolbow, C. Gerlach, On the completeness of meshfree particle methods. Int. J. Numer. Methods Eng. **43**(5), 785–819 (1998)

T. Belytschko, Y. Guo, W.K. Liu, S.P. Xiao, A unified stability analysis of meshless particle methods. Int. J. Numer. Methods Eng. **48**(9), 1359–1400 (2000)

J. Bonet, S.D. Kulasegaram, Correction and stabilization of smooth particle hydrodynamics methods with applications in metal forming simulations. Int. J. Numer. Methods Eng. **47**(6), 1189–1214 (2000)

J. Bonet, T.S.L. Lok, Variational and momentum preservation aspects of smooth particle hydrodynamic formulations. Comput. Methods Appl. Mech. Eng. **180**(1–2), 97–115 (1999)

L. Brookshaw, A method of calculating radiative heat diffusion in particle simulations. Publ. Astron. Soc. Aust. **6**(2), 207–210 (1985)

A. Colagrossi, M. Landrini, Numerical simulation of interfacial flows by smoothed particle hydrodynamics. J. Comput. Phys. **191**(2), 448–475 (2003)

S.J. Cummins, M. Rudman, An SPH projection method. J. Comput. Phys. **152**(2), 584–607 (1999)

C.T. Dyka, R.P. Ingel, An approach for tension instability in Smoothed Particle Hydrodynamics (SPH). Comput. Struct. **57**(4), 573–580 (1995)

C.T. Dyka, P.W. Randles, R.P. Ingel, Stress points for tension instability in SPH. Int. J. Numer. Methods Eng. **40**(13), 2325–2341 (1997)

P. Espanol, M. Revenga, Smoothed dissipative particle dynamics. Phys. Rev. E **67**(2), 026705 (2003)

M. Ferrand, D.R. Laurence, B.D. Rogers, D. Violeau, C. Kassiotis, Unified semi-analytical wall boundary conditions for inviscid, laminar or turbulent flows in the meshless SPH method. Int. J. Numer. Methods Fluids **71**(4), 446–472 (2013)

J.P. Fürstenau, H. Wessels, C. Weißenfels, P. Wriggers, Generating virtual process maps of SLM using powder scale SPH simulations. Comput. Part. Mech. **339**, 91–114 (2019)

G.C. Ganzenmüller, An hourglass control algorithm for Lagrangian Smooth Particle Hydrodynamics. Comput. Methods Appl. Mech. Eng. **286**, 87–106 (2015)

R.A. Gingold, J.J. Monaghan, Smoothed Particle Hydrodynamics: theory and application to non-spherical stars. Mon. Not. R. Astron. Soc. **181**, 375–389 (1977)

R.A. Gingold, J.J. Monaghan, Kernel estimates as a basis for general particle methods in hydrodynamics. J. Comput. Phys. **46**(3), 429–453 (1982)
X.Y. Hu, N.A. Adams, An incompressible multi-phase SPH method. J. Comput. Phys. **227**(1), 264–278 (2007)
S. Kulasegaram, J. Bonet, R.W. Lewis, M. Profit, A variational formulation based contact algorithm for rigid boundaries in two-dimensional SPH applications. Comput. Mech. **33**(4), 316–325 (2004)
S. Li, W.K. Liu, *Meshfree Particle Methods* (Springer, Berlin, Heidelberg, New York, 2007)
L.D. Libersky, A.G. Petschek, T.C. Carney, J.R. Hipp, F.A. Allahdadi, High strain Lagrangian hydrodynamics: a three-dimensional SPH code for dynamic material response. J. Comput. Phys. **109**(1), 67–75 (1993)
G.R. Liu, M.B. Liu, *Smoothed Particle Hydrodynamics: A Meshfree Particle Method* (World Scientific, 2003)
L.B. Lucy, Numerical approach to testing the fission hypothesis. Astron. J. **82**, 1013–1024 (1977)
J.R. MacDonald, Some simple isothermal equations of state. Rev. Mod. Phys. **38**(4), 669–679 (1966)
F. Macia, M. Antuono, L.M. González, A. Colagrossi, Theoretical analysis of the no-slip boundary condition enforcement in SPH methods. Prog. Theor. Phys. **125**(6), 1091–1121 (2011)
F. Macia, L.M. González, J.L. Cercos-Pita, A. Souto-Iglesias, A boundary integral SPH formulation: consistency and applications to ISPH and WCSPH. Prog. Theor. Phys. **128**(3), 439–462 (2012)
J.C. Marongiu, F. Leboeuf, J. Caro, E. Parkinson, Free surface flows simulations in Pelton turbines using an hybrid SPH-ALE method. J. Hydraul. Res. **48**(S1), 40–49 (2010)
S. Marrone, M.A.G.D. Antuono, A. Colagrossi, G. Colicchio, D. Le Touzé, G. Graziani, δ-SPH model for simulating violent impact flows. Comput. Methods Appl. Mech. Eng. **200**(13–16), 1526–1542 (2011)
D. Molteni, A. Colagrossi, A simple procedure to improve the pressure evaluation in hydrodynamic context using the SPH. Comput. Phys. Commun. **180**(6), 861–872 (2009)
J.J. Monaghan, An introduction to SPH. Comput. Phys. Commun. **48**(1), 89–96 (1988)
J.J. Monaghan, On the problem of penetration in particle methods. J. Comput. Phys. **82**(1), 1–15 (1989)
J.J. Monaghan, Smoothed Particle Hydrodynamics. Ann. Rev. Astron. Astrophys. **30**(1), 543–574 (1992)
J.J. Monaghan, Simulating free surface flows with SPH. J. Comput. Phys. **110**(2), 399–406 (1994)
J.J. Monaghan, SPH without a tensile instability. J. Comput. Phys. **159**(2), 290–311 (2000)
J.J. Monaghan, SPH compressible turbulence. Mon. Not. R. Astron. Soc. **335**(3), 843–852 (2002)
J.J. Monaghan, Smoothed Particle Hydrodynamics. Rep. Prog. Phys. **68**(8), 1703–1759 (2005)
J.J. Monaghan, R.A. Gingold, Shock simulation by the particle method SPH. J. Comput. Phys. **52**(2), 374–389 (1983)
J.J. Monaghan, J.B. Kajtar, SPH particle boundary forces for arbitrary boundaries. Comput. Phys. Commun. **180**(10), 1811–1820 (2009)
J.P. Morris, P.J. Fox, Y. Zhu, Modeling low reynolds number incompressible flows using SPH. J. Comput. Phys. **136**(1), 214–226 (1997)
R.P. Nelson, J.C.B. Papaloizou, Variable smoothing lengths and energy conservation in Smoothed Particle Hydrodynamics. Mon. Not. R. Astron. Soc. **270**(1), 1–20 (1994)
G. Oger, S. Marrone, D. Le Touzé, M. De Leffe, SPH accuracy improvement through the combination of a quasi-Lagrangian shifting transport velocity and consistent ALE formalisms. J. Comput. Phys. **313**, 76–98 (2016)
D.J. Price, Smoothed particle hydrodynamics and magnetohydrodynamics. J. Comput. Phys. **231**(3), 759–794 (2012)
D.J. Price, J.J. Monaghan, Smoothed particle magnetohydrodynamics–II. Variational principles and variable smoothing-length terms. Mon. Not. R. Astron. Soc. **348**(1), 139–152 (2004)
T. Rabczuk, T. Belytschko, S.P. Xiao, Stable particle methods based on Lagrangian kernels. Comput. Methods Appl. Mech. Eng. **193**(12–14), 1035–1063 (2004)
P.W. Randles, L.D. Libersky, Smoothed Particle Hydrodynamics: some recent improvements and applications. Comput. Methods Appl. Mech. Eng. **139**(1–4), 375–408 (1996)

P.W. Randles, L.D. Libersky, Normalized SPH with stress points. Int. J. Numer. Methods Eng. **48**(10), 1445–1462 (2000)

S. Shao, E.Y.M. Lo, Incompressible SPH method for simulating Newtonian and non-Newtonian flows with a free surface. Adv. Water Resour. **26**(7), 787–800 (2003)

M. Soleimani, *Numerical Simulation and Experimental Validation of Biofilm Formation*. Ph.D. thesis, Institut für Kontinuumsmechanik, Gottfried Wilhelm Leibniz Universität (2017)

J.W. Swegle, D.L. Hicks, S.W. Attaway, Smoothed Particle Hydrodynamics stability analysis. J. Comput. Phys. **116**(1), 123–134 (1995)

H. Takeda, S.M. Miyama, M. Sekiya, Numerical simulation of viscous flow by Smoothed Particle Hydrodynamics. Prog. Theor. Phys. **92**(5), 939–960 (1994)

A. Valizadeh, J.J. Monaghan, A study of solid wall models for weakly compressible SPH. J. Comput. Phys. **300**, 5–19 (2015)

R. Vignjevic, J. Campbell, L. Libersky, A treatment of zero-energy modes in the Smoothed Particle Hydrodynamics method. Comput. Methods Appl. Mech. Eng. **184**(1), 67–85 (2000)

J.P. Vila, On particle weighted methods and Smooth Particle Hydrodynamics. Math. Models Methods Appl. Sci. **9**(02), 161–209 (1999)

H. Wendland, Piecewise polynomial, positive definite and compactly supported radial functions of minimal degree. Adv. Comput. Math. **4**(1), 389–396 (1995)

R. Xu, P. Stansby, D. Laurence, Accuracy and stability in incompressible SPH (ISPH) based on the projection method and a new approach. J. Comput. Phys. **228**(18), 6703–6725 (2009)

Chapter 7
Peridynamics

Compared to SPH, Peridynamics is a more recent meshfree method. It is understood as the application of Molecular Dynamics to the macroscopic description of continua. Analogous to Molecular Dynamics (see, e.g., Griebel et al. 2013), Peridynamics distinguishes between bond-based and state-based formulations. In the former (Silling 2000), the force results solely from the displacements between two particles. In the state-based approach (Silling et al. 2007), the force at a particle is determined from the deformations of all particles in the neighborhood. State-based Peridynamics further divides into an ordinary and a non-ordinary approach. In the former, the line of action of the stress between two particles is assumed to run along with its distance. In non-ordinary Peridynamics, this restriction is removed. Moreover, the classical continuum mechanical material laws can be used employing the so-called correspondence formulation (Fig. 7.1). This work is limited solely to this concept. Descriptions and applications of alternative approaches can be found, for example, in Bobaru et al. (2016).

7.1 Correspondence Formulation

The discretization concept of Peridynamics can be assigned to the one-point methods. The equations are usually based on the total Lagrangian description and formulated with respect to the position vector $\mathbf{X}$ of the initial configuration. As will be shown in this section, the equations of Peridynamics can be derived directly from the weak form.

C. Weißenfels, *Simulation of Additive Manufacturing using Meshfree Methods*,
Lecture Notes in Applied and Computational Mechanics 97,
https://doi.org/10.1007/978-3-030-87337-0_7

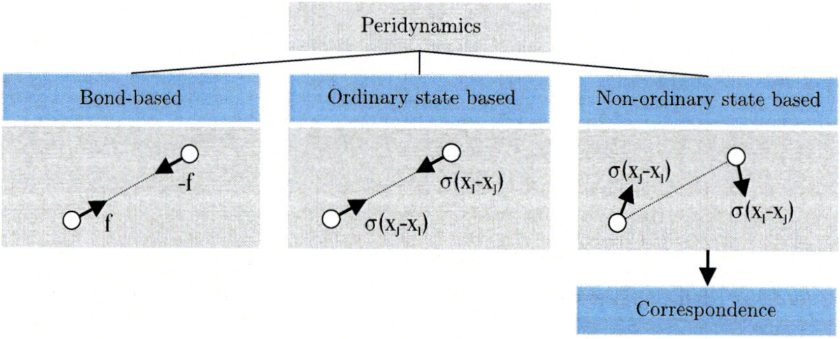

Fig. 7.1 Classification of different discretization concepts in Peridynamics

7.1.1 Theory

State-based Peridynamics, formulated in Silling et al. (2007), replaces the divergence of stress in the balance of momentum by the difference of force vector state fields $\underline{\mathbf{T}}$ integrated over the neighborhood $\mathcal{H}_{\mathbf{X}}$ of $\mathbf{X}$

$$\rho_0 \ddot{\mathbf{u}}(\mathbf{X}, t) = \int_{\mathcal{H}_{\mathbf{X}}} \left[\underline{\mathbf{T}}\langle \mathbf{X}' - \mathbf{X} \rangle - \underline{\mathbf{T}}'\langle \mathbf{X} - \mathbf{X}' \rangle \right] \, \mathrm{d}V_{\mathbf{X}'} + \rho_0 \bar{\mathbf{b}}. \tag{7.1}$$

The geometry of $\mathcal{H}_{\mathbf{X}}$ is assumed to be spherical. The radius is called the horizon and abbreviated with the symbol δ. The family of $\mathbf{X}$ includes all bonds within $\mathcal{H}_{\mathbf{X}}$. A bond corresponds to the difference $\mathbf{X}' - \mathbf{X}$, where $\mathbf{X}'$ denotes a mass point in the neighborhood of $\mathbf{X}$. Peridynamics also introduces the notion of states. This term, borrowed from thermodynamics, describes the mapping of bonds and is denoted by an underscore. The bond to be mapped is written inside an angle bracket. Thus, the force vector state field $\underline{\mathbf{T}}$ corresponds to the mapping that assigns a bond of the family of $\mathbf{X}$ a vector with unit [N/m^6]. The force vector state field $\underline{\mathbf{T}}'$ describes the mapping of bonds of the family of $\mathbf{X}'$. The correspondence formulation is based on the introduction of the deformation vector state field $\underline{\mathbf{Y}}$. This state field maps the bond into the current configuration

$$\underline{\mathbf{Y}}\langle \mathbf{X}' - \mathbf{X} \rangle = \mathbf{x}' - \mathbf{x}. \tag{7.2}$$

The equivalence with the deformation gradient (3.5) is given if both mappings lead to the same result. Then it follows that the strain energy density of the peridynamic formulation V is equal to the corresponding continuum mechanical quantity W

$$\underline{\mathbf{Y}}\langle \mathbf{X}' - \mathbf{X} \rangle = \mathbf{F}(\mathbf{X}' - \mathbf{X}) \quad \rightarrow \quad V(\underline{\mathbf{Y}}) = W(\mathbf{F}) = \rho_0 \psi(\mathbf{F}). \tag{7.3}$$

In Silling et al. (2007), a reduction operator $\mathcal{R}$ is introduced. Using this description, an approximated $\mathbf{F}$ can be defined that results directly from the deformation vector state field

$$\begin{aligned} \mathbf{F}(\underline{\mathbf{Y}}) &= \mathcal{R}\{\underline{\mathbf{Y}}\} = \left(\underline{\mathbf{Y}} * \underline{\mathbf{X}}\right)\mathbf{K}^{-1} \\ &= \int_{\mathcal{H}_\mathbf{X}} \omega\left(\mathbf{X}' - \mathbf{X}\right)\left(\mathbf{x}' - \mathbf{x}\right) \otimes \left(\mathbf{X}' - \mathbf{X}\right)\, dV_{\mathbf{X}'}\mathbf{K}^{-1}\left(\mathbf{X}\right). \end{aligned} \tag{7.4}$$

The operator $*$ corresponds to the tensor product of states and ω is indicated as the influence function. An exact definition and more information on mathematical rules in Peridynamics can be found in Silling et al. (2007). The shape tensor $\mathbf{K}$ is determined from the tensor product of the reference position vector state $\underline{\mathbf{X}}$ with itself

$$\mathbf{K} = \underline{\mathbf{X}} * \underline{\mathbf{X}} = \int_{H_\mathbf{X}} \omega\left(\mathbf{X}' - \mathbf{X}\right)\left(\mathbf{X}' - \mathbf{X}\right) \otimes \left(\mathbf{X}' - \mathbf{X}\right)\, dV_{\mathbf{X}'}. \tag{7.5}$$

The 1st Piola Kirchhoff stress tensor $\mathbf{P}$ is calculated for hyperelastic material behavior from the derivative of the strain energy density with respect to the deformation gradient

$$\dot{W} = \frac{\partial W}{\partial \mathbf{F}} \cdot \dot{\mathbf{F}}(\underline{\mathbf{Y}}) = \mathbf{P} \cdot \dot{\mathbf{F}}(\underline{\mathbf{Y}}). \tag{7.6}$$

Considering (7.4) and exploiting the symmetry of the shape tensor $\mathbf{K}$, the right-hand side can be reformulated

$$\begin{aligned} \mathbf{P} \cdot \dot{\mathbf{F}}(\underline{\mathbf{Y}}) &= \int_{\mathcal{H}_\mathbf{X}} \omega\left(\mathbf{X}' - \mathbf{X}\right)\mathbf{P}\left(\mathbf{X}\right) \cdot \left(\dot{\mathbf{x}}' - \mathbf{x}\right) \otimes \left(\mathbf{X}' - \mathbf{X}\right)\, dV_{\mathbf{X}'}\mathbf{K}^{-1}\left(\mathbf{X}\right) \\ &= \int_{\mathcal{H}_\mathbf{X}} \omega\left(\mathbf{X}' - \mathbf{X}\right)\mathbf{P}\left(\mathbf{X}\right)\mathbf{K}^{-1}\left(\mathbf{X}\right)\left(\mathbf{X}' - \mathbf{X}\right) \cdot \left(\dot{\mathbf{x}}' - \dot{\mathbf{x}}\right)\, dV_{\mathbf{X}'}. \end{aligned} \tag{7.7}$$

By considering the correspondence (7.3) the derivative of W with respect to time must be equal to $\dot{V}$. Substituting (7.2) into (7.6) establishes a relation between $\mathbf{P}$ and $\underline{\mathbf{T}}$

$$\dot{V} = \frac{\partial V}{\partial \underline{\mathbf{Y}}} \bullet \dot{\underline{\mathbf{Y}}} = \underline{\mathbf{T}} \bullet \dot{\underline{\mathbf{Y}}} = \int_{\mathcal{H}_\mathbf{X}} \underline{\mathbf{T}} \cdot \left(\dot{\mathbf{x}}' - \dot{\mathbf{x}}\right)\, dV_{\mathbf{X}'} = \mathbf{P} \cdot \dot{\mathbf{F}}(\underline{\mathbf{Y}}). \tag{7.8}$$

The operator $\bullet$ denotes the scalar product of two states. The comparison between (7.8) and (7.7) provides the relationship between the 1st Piola Krichhoff stress tensor and the force vector state field

$$\underline{\mathbf{T}}\langle \mathbf{X}' - \mathbf{X}\rangle = \omega\left(\mathbf{X}' - \mathbf{X}\right)\mathbf{P}\left(\mathbf{X}\right)\mathbf{K}^{-1}\left(\mathbf{X}\right)\left(\mathbf{X}' - \mathbf{X}\right). \tag{7.9}$$

Remark 1 The deformation vector state field corresponds to a non-local measure, since only finite differences are considered. In contrast, the deformation gradient describes a local behavior. The relation between $\underline{\mathbf{Y}}$ and $\mathbf{F}$ follows from a Taylor series (Silling and Lehoucq 2010). The current position vector in the neighborhood of the point P identified by $\mathbf{X}$ can also be represented in terms of derivatives

$$\mathbf{x}' = \mathbf{x}\left(\mathbf{X}'\right) = \mathbf{x}\left(\mathbf{X}\right) + \frac{\partial \mathbf{x}}{\partial \mathbf{X}}\left[\mathbf{X}' - \mathbf{X}\right] + \mathcal{O}\left(\left[\mathbf{X}' - \mathbf{X}\right]^2\right). \tag{7.10}$$

Using (3.5) and exploiting (7.2) yields to the relationship between the deformation vector state field and the deformation gradient

$$\underline{\mathbf{Y}}\langle\mathbf{X}' - \mathbf{X}\rangle = \mathbf{x}\left(\mathbf{X}'\right) - \mathbf{x}\left(\mathbf{X}\right) = \mathbf{F}\left(\mathbf{X}' - \mathbf{X}\right) + \mathcal{O}\left(\left[\mathbf{X}' - \mathbf{X}\right]^2\right). \tag{7.11}$$

In the limit as δ approaches zero, i.e. $\delta \rightarrow 0$, the higher order contributions can be neglected and $\underline{\mathbf{Y}}$ equals $\mathbf{F}$.

Remark 2 To show the analogy and the differences between the descriptions in Chap. 3 and the peridynamic formulations, the continuum mechanical abbreviations are given. In Silling et al. (2007), the peridynamic strain energy density is indicated by the letter W. In addition, the position vector of the initial configuration is abbreviated by $\mathbf{x}$ and of the current configuration by $\mathbf{y}$.

7.1.2 Relation to Weak Form

Analogous to Sect. 4.2.4, the weak form can be represented in terms of differences with respect to the initial configuration. With the definition of the vector $\mathbf{T}$

$$\mathbf{T}\left(\mathbf{X}, \mathbf{X}'\right) = \omega\left(\mathbf{X}' - \mathbf{X}\right)\mathbf{P}(\mathbf{X})\mathbf{M}^{-1}\left(\mathbf{X}\right)\left[\mathbf{X}' - \mathbf{X}\right] \tag{7.12}$$

the virtual internal work can be formulated in dependence on the position vector $\mathbf{X}$

$$\int_{B_t} \operatorname{Grad} \boldsymbol{\eta} \cdot \mathbf{P}\, \mathrm{d}v = -\int_{B_0}\int_{H_\mathbf{X}} \boldsymbol{\eta}\left(\mathbf{X}\right) \cdot \left[\mathbf{T}\left(\mathbf{X}, \mathbf{X}'\right) - \mathbf{T}\left(\mathbf{X}', \mathbf{X}\right)\right]\, \mathrm{d}V_{\mathbf{X}'}\, \mathrm{d}V_{\mathbf{X}}. \tag{7.13}$$

The moment tensor of the initial configuration can be identified as the shape tensor $\mathbf{K}$

$$\mathbf{M}\left(\mathbf{X}\right) = \int_{H_\mathbf{X}} \omega\left(\mathbf{X}' - \mathbf{X}\right)\left(\mathbf{X}' - \mathbf{X}\right) \otimes \left(\mathbf{X}' - \mathbf{X}\right)\, \mathrm{d}V_{\mathbf{X}'} = \mathbf{K}\left(\mathbf{X}\right). \tag{7.14}$$

Thus, the vector $\mathbf{T}$ from (7.12) corresponds exactly to the force vector state field (7.9)

$$\mathbf{T}\left(\mathbf{X}, \mathbf{X}'\right) = \underline{\mathbf{T}}\langle\mathbf{X}' - \mathbf{X}\rangle. \tag{7.15}$$

Exploiting (7.13) and (7.15) the weak form with respect to the initial configuration can be formulated in dependence on $\underline{\mathbf{T}}$

$$
\begin{aligned}
&-\int_{B_0}\int_{H_{\mathbf{X}}} \boldsymbol{\eta}\cdot\left[\underline{\mathbf{T}}\langle\mathbf{X}'-\mathbf{X}\rangle-\underline{\mathbf{T}}'\langle\mathbf{X}-\mathbf{X}'\rangle\right]\,\mathrm{d}V_{\mathbf{X}'}\,\mathrm{d}V_{\mathbf{x}}+\int_{B_0}\rho_0\boldsymbol{\eta}\cdot\left(\ddot{\mathbf{u}}-\bar{\mathbf{b}}\right)\,\mathrm{d}V\\
&=\int_{\partial B_0^{N_{\mathbf{u}}}}\boldsymbol{\eta}\cdot\bar{\mathbf{T}}\,\mathrm{d}A.
\end{aligned}
\tag{7.16}
$$

Since the test function can be arbitrary and thus non-zero, the above equation only holds, if (7.1) is fulfilled

$$
\rho_0\ddot{\mathbf{u}}(\mathbf{X},t)=\int_{H_{\mathbf{X}}}\left[\underline{\mathbf{T}}\langle\mathbf{X}'-\mathbf{X}\rangle-\underline{\mathbf{T}}'\langle\mathbf{X}-\mathbf{X}'\rangle\right]\,\mathrm{d}V+\rho_0\bar{\mathbf{b}}.
$$

However, this relation only applies to inner mass points. For mass points with $\bar{\mathbf{T}}\neq 0$, the Neumann boundary condition must be taken into account due to (7.16).

7.1.3 Discretization

Peridynamics, like Smoothed Particle Hydrodynamics, is a one-point methods that subdivides the entire domain into individual particles. In the discrete form, the integrals are replaced by a summation over all particles n and over all particles N_I in the neighborhood of particle I. For inner particles, (7.16) or (7.1) directly yields to the discrete relation

$$
\sum_{I=1}^{n}\left\{-\sum_{J=1}^{N_I}\left[\underline{\mathbf{T}_I}\langle\mathbf{X}_J-\mathbf{X}_I\rangle-\underline{\mathbf{T}_J}\langle\mathbf{X}_I-\mathbf{X}_J\rangle\right]V_J+\rho_0\left[\ddot{\mathbf{u}}\left(\mathbf{X}_I\right)-\bar{\mathbf{b}}\right]\right\}V_I=0
\tag{7.17}
$$

The discretization of the force vector state field follows from (7.9). Due to the symmetry of the shape tensor, the relation can be reformulated

$$
\underline{\mathbf{T}_I}\langle\mathbf{X}_J-\mathbf{X}_I\rangle=\omega\left(\mathbf{X}_J-\mathbf{X}_I\right)\mathbf{P}_I\mathbf{K}_I^{-1}\left[\mathbf{X}_J-\mathbf{X}_I\right]=\mathbf{P}_I\omega\left(\mathbf{X}_J-\mathbf{X}_I\right)\left[\mathbf{X}_J-\mathbf{X}_I\right]\mathbf{K}_I^{-1}.
\tag{7.18}
$$

The subscript indicates the evaluation of a quantity at the corresponding particle, e.g.

$$
\mathbf{P}_I:=\mathbf{P}\left(\mathbf{X}_I\right),\quad \mathbf{K}_I:=\mathbf{K}\left(\mathbf{X}_I\right).
\tag{7.19}
$$

The shape tensor at the particle is obtained from (7.5) by summing over all particles in the neighborhood

$$
\mathbf{K}_I=\sum_{J=1}^{N_I}\omega\left(\mathbf{X}_J-\mathbf{X}_I\right)\left(\mathbf{X}_J-\mathbf{X}_I\right)\otimes\left(\mathbf{X}_J-\mathbf{X}_I\right)V_J.
\tag{7.20}
$$

By substituting (7.18), the discrete form of the balance equation (7.17) can be formulated in dependence on the 1st Piola Kirchhoff stress tensor. The internal force due to the state of stress in the material is given by

$$\mathbf{f}_I^s = -\sum_{J=1}^{N_I} \Big[\omega\left(\mathbf{X}_J - \mathbf{X}_I\right)\mathbf{P}_I\mathbf{K}_I^{-1}\left[\mathbf{X}_J - \mathbf{X}_I\right] - \omega\left(\mathbf{X}_I - \mathbf{X}_J\right)\mathbf{P}_J\mathbf{K}_J^{-1}\left[\mathbf{X}_I - \mathbf{X}_J\right]\Big] V_J V_I. \tag{7.21}$$

In its original form, Peridynamics does not use shape functions. However, the influence function ω multiplied by the distance can be interpreted as the derivative of a weighting function

$$\frac{\partial W(\mathbf{X}_I - \mathbf{X}_J)}{\partial \mathbf{X}_I} = \omega\left(\mathbf{X}_I - \mathbf{X}_J\right)\left[\mathbf{X}_I - \mathbf{X}_J\right] \tag{7.22}$$

using the definition

$$\omega\left(\mathbf{X}_I - \mathbf{X}_J\right) = \frac{\partial W(\mathbf{X}_I - \mathbf{X}_J)}{\partial \|\mathbf{X}_I - \mathbf{X}_J\|} \frac{1}{\|\mathbf{X}_I - \mathbf{X}_J\|}. \tag{7.23}$$

Equation (7.18) can thus be alternatively formulated in dependence on weighting functions

$$\underline{\mathbf{T}}_I\langle\mathbf{X}_J - \mathbf{X}_I\rangle = \mathbf{P}_I \frac{\partial W\left(\mathbf{X}_J - \mathbf{X}_I\right)}{\partial \mathbf{X}_J}\mathbf{K}_I^{-1}. \tag{7.24}$$

With the above relation, the internal force $\mathbf{f}_I^s$ due to the state of stress in the material can also be written as

$$\mathbf{f}_I^s = -\sum_{J=1}^{N_I}\left[\mathbf{P}_I \frac{\partial W\left(\mathbf{X}_J - \mathbf{X}_I\right)}{\partial \mathbf{X}_J}\mathbf{K}_I^{-1} - \mathbf{P}_J \frac{\partial W\left(\mathbf{X}_I - \mathbf{X}_J\right)}{\partial \mathbf{X}_I}\mathbf{K}_J^{-1}\right] V_I V_J. \tag{7.25}$$

The contributions from the inertia term and the dead load are determined analogously to (4.16). The calculation of the stress requires the deformation gradient. The discrete form follows directly from (7.4) and can also be formulated on the basis of weighting functions

$$\begin{aligned}\mathbf{F}_I &= \sum_{J=1}^{N_I} \omega\left(\mathbf{X}_J - \mathbf{X}_I\right)\left(\mathbf{x}_J - \mathbf{x}_I\right) \otimes \left(\mathbf{X}_J - \mathbf{X}_I\right) V_J \mathbf{K}_I^{-1} \\ &= \sum_{J=1}^{N_I} \left(\mathbf{x}_J - \mathbf{x}_I\right) \otimes \frac{\partial W\left(\mathbf{X}_J - \mathbf{X}_I\right)}{\partial \mathbf{X}_J}\mathbf{K}_I^{-1} V_J.\end{aligned} \tag{7.26}$$

Since the equations within Peridynamics are formulated in their original form in the total Lagrangian description, the phenomenon of tensile instability does not occur.

7.2 Reproducing Conditions

Taking advantage of (7.22), the shape tensor (7.20) can also be calculated in terms of weighting functions. Exploiting (6.12) shows that **K** corresponds exactly to the negative correction tensor **L** (6.44) formulated with respect to the initial configuration, see also (Ganzenmüller et al. 2015)

$$\begin{aligned}\mathbf{K}_I &= \sum_{J=1}^{N_I} (\mathbf{X}_J - \mathbf{X}_I) \otimes \frac{\partial W(\mathbf{X}_J - \mathbf{X}_I)}{\partial \mathbf{X}_J} V_J \\ &= -\sum_{J=1}^{N_I} (\mathbf{X}_J - \mathbf{X}_I) \otimes \frac{\partial W(\mathbf{X}_I - \mathbf{X}_J)}{\partial \mathbf{X}_I} V_J = -\mathbf{L}(\mathbf{X}_I)\,. \end{aligned} \tag{7.27}$$

The prerequisite to satisfy the 1st order reproducing condition in the derivatives is the fulfillment of the corresponding requirement for the 0th order

$$\sum_{J=1}^{N_I} \frac{\partial W(\mathbf{X}_I - \mathbf{X}_J)}{\partial \mathbf{X}_I} V_J = \sum_{J=1}^{N_I} \omega(\mathbf{X}_I - \mathbf{X}_J)\,[\mathbf{X}_I - \mathbf{X}_J]\, V_J = 0. \tag{7.28}$$

7.3 Discrete Conservation Properties

As a consequence of (7.27), and taking advantage of (6.12), (7.25) can also be formulated with respect to the correction tensor **L**

$$\mathbf{f}_I^s = -\sum_{J=1}^{N_I} \left[\mathbf{P}_I \frac{\partial W(\mathbf{X}_I - \mathbf{X}_J)}{\partial \mathbf{X}_I} \mathbf{L}_I^{-1} - \mathbf{P}_J \frac{\partial W(\mathbf{X}_J - \mathbf{X}_I)}{\partial \mathbf{X}_J} \mathbf{L}_J^{-1} \right] V_I V_J. \tag{7.29}$$

Hence, the peridynamic equations lead to the same formulations as in the SPH, if the divergence of stress is based on the quotient rule (6.70). The only difference is the selection of the weighting function, which follows according to (7.22) in standard Peridynamics. As pointed out in Sect. 6.6, the global balance of momentum is automatically conserved, since the calculation can be represented as the difference of forces between particles

$$\mathbf{f}_I^s = \sum_{J=1}^{N_I} [\mathbf{f}_{IJ} - \mathbf{f}_{JI}]\,, \quad \mathbf{f}_{IJ} = \mathbf{P}_I \frac{\partial W(\mathbf{X}_I - \mathbf{X}_J)}{\partial \mathbf{X}_I} \mathbf{L}_I^{-1} V_I V_J\,. \tag{7.30}$$

However, the requirement on the global balance of angular momentum is not automatically fulfilled, see also Sect. 6.6.

Remark Approaches based on alternative non-local derivative rules lead to a force calculation based only on the shape tensor at the particle under consideration (Madenci et al. 2018; Tu and Li 2017)

$$\mathbf{f}_I^s = -\sum_{J=1}^{N_I} \omega\left(\mathbf{X}_J - \mathbf{X}_I\right)\left[\mathbf{P}_J - \mathbf{P}_I\right]\mathbf{K}_I^{-1}\left(\mathbf{X}_J - \mathbf{X}_I\right) V_I V_J. \tag{7.31}$$

Taking advantage of (7.22), (7.27), (6.12), and of the symmetry of $\mathbf{K}$, the calculation can be related to the weighting function and the correction tensor $\mathbf{L}$

$$\begin{aligned} \mathbf{f}_I^s &= -\sum_{J=1}^{N_I} \left[\mathbf{P}_J - \mathbf{P}_I\right] \frac{\partial W\left(\mathbf{X}_J - \mathbf{X}_I\right)}{\partial \mathbf{X}_J} \mathbf{K}_I^{-1} V_I V_J \\ &= -\sum_{J=1}^{N_I} \left[\mathbf{P}_J - \mathbf{P}_I\right] \frac{\partial W\left(\mathbf{X}_I - \mathbf{X}_J\right)}{\partial \mathbf{X}_I} \mathbf{L}_I^{-1} V_I V_J. \end{aligned} \tag{7.32}$$

The comparison with (6.46) shows that this variant corresponds to the approach in the SPH which is based on the product rule.

7.4 Integration Constraint

Since the equations in Peridynamics can be derived from the weak form, the integration constraint must be satisfied. To pass the standard patch test, for a constant stress, i.e. $\mathbf{P} := \bar{\mathbf{P}} = const$, it has to hold

$$\int_{B_0} \operatorname{Grad} \eta \cdot \bar{\mathbf{P}} \, \mathrm{d}V_{\mathbf{X}} = \int_{\partial B_0} \eta \cdot \bar{\mathbf{P}}\mathbf{N} \, \mathrm{d}A_{\mathbf{X}}. \tag{7.33}$$

Exploiting (7.13) and (7.9) results in the necessary requirement in the framework of Peridynamics

$$\begin{aligned} &-\int_{B_0}\int_{H_{\mathbf{X}}} \Big[\omega\left(\mathbf{X}' - \mathbf{X}\right)\mathbf{K}^{-1}\left(\mathbf{X}\right)\left(\mathbf{X}' - \mathbf{X}\right) \\ &- \omega\left(\mathbf{X} - \mathbf{X}'\right)\mathbf{K}^{-1}\left(\mathbf{X}'\right)\left(\mathbf{X} - \mathbf{X}'\right)\Big] \, \mathrm{d}V_{\mathbf{X}'} \, \mathrm{d}V_{\mathbf{X}} = \int_{\partial B_0} \mathbf{N} \, \mathrm{d}A_{\mathbf{X}}. \end{aligned} \tag{7.34}$$

The discrete case leads to a condition that has to be fulfilled at each particle

$$
\begin{aligned}
&- \sum_{J=1}^{N_I} \Big[\omega \left(\mathbf{X}_J - \mathbf{X}_I\right) \mathbf{K}_I^{-1} \left(\mathbf{X}_J - \mathbf{X}_I\right) \\
&- \omega \left(\mathbf{X}_I - \mathbf{X}_J\right) \mathbf{K}_J^{-1} \left(\mathbf{X}_I - \mathbf{X}_J\right) \Big] V_I V_J = \begin{cases} 0 & \text{inner particles} \\ \bar{\mathbf{N}}_I & \text{boundary particles} \end{cases} .
\end{aligned}
\tag{7.35}
$$

The resulting normal vector $\bar{\mathbf{N}}_I$ is determined from the discrete right-hand side in (7.34).

Remark Since $\mathbf{N}_I$ is difficult to determine at the boundary, a normal vector can be defined beforehand. The shape or weighting functions must then be determined such that the integration constraint is always satisfied. A corresponding approach can be found in Bode et al. (2021).

7.5 Search Algorithm

The neighborhood in Peridynamics is usually spherical. All particles whose distance to $\mathbf{X}_I$ is smaller than a given value δ are part of the neighborhood H_I of particle I

$$
H_I := \{\mathbf{X}_J \in B_0 \mid \|\mathbf{X}_J - \mathbf{X}_I\| < \delta\} . \tag{7.36}
$$

In standard Peridynamics, the equations are formulated with respect to the initial configuration. The neighborhood is defined beforehand and is not changed during the calculation. The requirements on the search algorithm for maintaining the equilibrium of forces during the calculation are therefore no longer necessary. The particle distribution at the beginning of the calculation is usually grid-shaped and therefore statistically isotropic and homogeneous. The only remaining requirement is the correct assignment of force pairs. If particle J is part of the neighborhood of I, particle I must also be part of the neighborhood of J. If the search radii are the same, this condition is always satisfied.

With a varying search radius $\delta := \delta(\mathbf{X}_I)$ this condition can be violated (Fig. 7.2). Thus, the global conservation of momentum is not automatically fulfilled and unphysical forces occur. In Silling et al. (2015) these are named ghost forces and the introduction of an additional partial stress is recommended to counteract unphysical outcomes. Another approach recommends the application of two different neighbor-

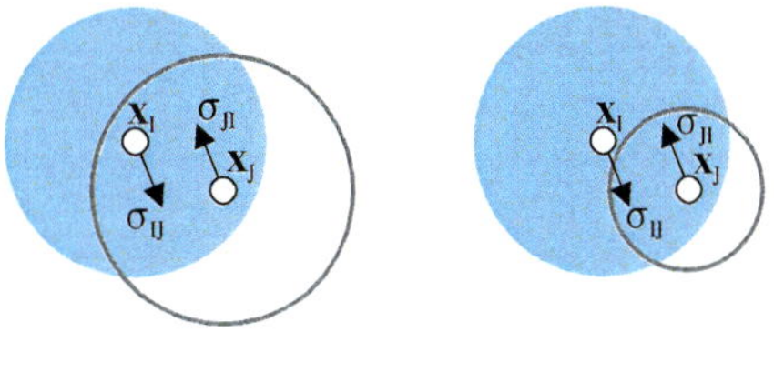

Fig. 7.2 (Left) Fulfillment of the criterion on the search algorithm due to constant search radii. (Right) Violation of the constraint due to different search radii

hoods to meet the requirements for the global conservation of momentum in this case (Ren et al. 2016).

7.6 Configurational Consistency

Two approaches are developed to avoid zero energy modes, but actually these schemes ensure the configurational consistency. Breitzman and Dayal (2018) recommends to calculate a separate deformation gradient at each bond in the neighborhood instead of just evaluating **F** at **X**

$$\mathbf{F}^{bond}\left(\mathbf{X},\mathbf{X}'\right)=\mathbf{F}\left(\mathbf{X}\right)-\frac{1}{\|\mathbf{X}'-\mathbf{X}\|^2}\left(\left[\mathbf{F}\left(\mathbf{X}\right)\left(\mathbf{X}'-\mathbf{X}\right)-\left(\mathbf{x}'-\mathbf{x}\right)\right]\otimes\left(\mathbf{X}'-\mathbf{X}\right)\right). \tag{7.37}$$

The violation of the configurational consistency (4.84) is subtracted from the original deformation gradient **F**. If this deviation equals zero, (7.37) results in the standard formulation of the deformation gradient. In Littlewood (2010) and Silling (2017) the violation of the configurational consistency is corrected by a penalty approach. The force vector state field is enhanced by an additional contribution similar to the OTM method (5.29), or the SPH (6.80)

$$\underline{\mathbf{T}}^{stab}\langle\mathbf{X}'-\mathbf{X}\rangle=\underline{\mathbf{T}}\langle\mathbf{X}'-\mathbf{X}\rangle+c\frac{\mathbf{x}'-\mathbf{x}}{\|\mathbf{x}'-\mathbf{x}\|}\frac{\|\mathbf{x}'-\mathbf{x}\|-\|\mathbf{F}\left(\mathbf{X}'-\mathbf{X}\right)\|}{\|\mathbf{X}'-\mathbf{X}\|}. \tag{7.38}$$

The parameter c corresponds to the penalty parameter.

7.7 Rank Instability

As shown in the previous sections, the correspondence formulation is equivalent to a variant of the SPH. Thus, the negative effects due to under-integration, which are described in Sect. 6.13, also occur in Peridynamics. No energy can be assigned to individual deformation states. Unphysical overlaps and interpenetrations can result (Tupek and Radovitzky 2014).

To eliminate these negative effects, different approaches are developed in the context of Peridynamics. A comparison of some stabilization methods can be found in Breitenfeld et al. (2014). However, the concepts from Sect. 6.13 can also be applied.

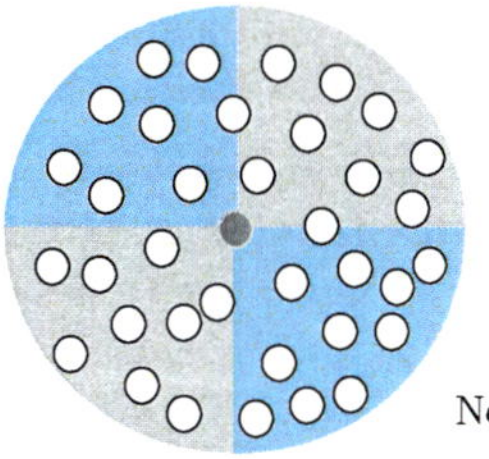

Fig. 7.3 Subdivision of the neighborhood into sub-neighborhoods

7.7.1 Alternative Formulations

Tupek and Radovitzky (2014) proposes an alternative strain measure based on single bonds. Chowdhury et al. (2019) suggests the evaluation of the peridynamic quantities in sub-neighborhoods to eliminate the negative effects due to under-integration (Fig. 7.3). The force at the particle (7.25) is thus calculated from the sum over individual subareas N_I^{sub}

$$\mathbf{f}_I^s = -\sum_{\alpha=1}^{N_I^{sub}} \sum_{J=1}^{N_I^\alpha} \left[\mathbf{P}_I^\alpha \frac{\partial W\left(\mathbf{X}_J - \mathbf{X}_I\right)}{\partial \mathbf{X}_J} \left(\mathbf{K}_I^\alpha\right)^{-1} - \mathbf{P}_J^\alpha \frac{\partial W\left(\mathbf{X}_I - \mathbf{X}_J\right)}{\partial \mathbf{X}_I} \left(\mathbf{K}_J^\alpha\right)^{-1} \right] V_I V_J. \tag{7.39}$$

In each sub-neighborhood, the calculated kinematic or constitutive quantity can be different. This increases the number of evaluation points in the neighborhood. As shown in Hartmann et al. (2020) this concept can lead to unphysical outcomes in the case of large deformations.

7.7.2 Generalization of Peridynamics

In classical Peridynamics, the quantities are evaluated only at the central particle in the respective neighborhood. This reduces the number of evaluation points. In the generalized form introduced in Bode et al. (2020a), this limitation is removed. The individual quantities are determined separately at each particle in the neighborhood.

As already presented in Sect. 7.1.3, the discrete balance equation as well as the calculation of the deformation gradient can be described in dependence on radial weighting functions W. As shown in (6.6), there exists a relationship between shape functions and W. Thus, the deformation gradient at any particle in the neighborhood can also be calculated from the derivative of freely chosen shape functions (Bode et al. 2020a)

$$\mathbf{F}^I\left(\mathbf{X}_J\right) = \sum_{K=1}^{N_I} \left(\mathbf{x}_K - \mathbf{x}_I\right) \otimes \frac{\partial N_K^I\left(\mathbf{X}_J\right)}{\partial \mathbf{X}_J}. \tag{7.40}$$

The superscript denotes the neighborhood, i,e. $\mathbf{F}^I(\mathbf{X}_J)$ corresponds to the deformation gradient at the position $\mathbf{X}_J$ in the neighborhood of particle I. The generalization leads to an alternative discretization of the force vector state field. First, the stress at each particle in the neighborhood is now different, and second, the derivative is approximated by shape functions

$$\underline{\mathbf{T}_I}\langle \mathbf{X}_J - \mathbf{X}_I\rangle = -\frac{1}{V_I}\mathbf{P}^I(\mathbf{X}_J)\,\frac{\partial N_I^J(\mathbf{X}_J)}{\partial \mathbf{X}_J}. \tag{7.41}$$

Thus, the calculation for the internal force due to the state of stress in the material changes to

$$\mathbf{f}_I^s = \sum_{J=1}^{N_I}\left[\frac{1}{V_I}\mathbf{P}^I(\mathbf{X}_J)\,\frac{\partial N_I^J(\mathbf{X}_J)}{\partial \mathbf{X}_J} - \frac{1}{V_J}\mathbf{P}^J(\mathbf{X}_I)\,\frac{\partial N_J^I(\mathbf{X}_I)}{\partial \mathbf{X}_I}\right]V_J V_I. \tag{7.42}$$

The transition to the classical correspondence formulation in Peridynamics results from the relations

$$\begin{aligned}\frac{\partial N_K^I(\mathbf{X}_J)}{\partial \mathbf{X}_J} &= \omega(\mathbf{X}_K - \mathbf{X}_I)\,V_K\mathbf{K}_I^{-1}(\mathbf{X}_K - \mathbf{X}_I)\\ \frac{\partial N_J^I(\mathbf{X}_I)}{\partial \mathbf{X}_I} &= \omega(\mathbf{X}_I - \mathbf{X}_J)\,V_J\mathbf{K}_J^{-1}(\mathbf{X}_J - \mathbf{X}_I)\\ \frac{\partial N_I^J(\mathbf{X}_J)}{\partial \mathbf{X}_J} &= \omega(\mathbf{X}_J - \mathbf{X}_I)\,V_I\mathbf{K}_I^{-1}(\mathbf{X}_I - \mathbf{X}_J).\end{aligned} \tag{7.43}$$

By substituting (7.43) into (7.40), it can be directly seen that the deformation gradient is evaluated only at the central particle, i.e. $\mathbf{F}^I(\mathbf{X}_J) \to \mathbf{F}_I$. The same is thus true for the stress $\mathbf{P}^I(\mathbf{X}_J) \to \mathbf{P}_I$.

As shown in Bode et al. (2020a), no negative effects due to under-integration occur employing this approach, provided that shape functions with a sufficiently high order are used. Moreover, no additional numerical parameters are needed. An extension to incompressible material behavior can be found in Bode et al. (2020b).

7.8 Boundary Conditions

Similar to SPH, Peridynamics differs from Galerkin methods regarding the imposition of boundary conditions. Mostly, the concept of ghost particles is employed. The number and the position of the ghost particles are chosen such that the outer particles have a complete neighborhood (Macek and Silling 2007; Littlewood 2015). The Dirichlet boundary conditions are defined at the ghost particles. In Silling and Askari (2014) a linear distribution of these values over the position of the ghost particles is proposed. An alternative approach uses a correction of the shape function at the outer layer to apply the displacements directly at the boundaries (Wu 2014). So-called general meshfree approximation functions (Wu et al. 2011) are introduced, which possess a weak Kronecker-δ property. Neumann boundary conditions are converted into corresponding volume forces (Madenci et al. 2018).

The generalized approach from Bode et al. (2020a) also offers a big advantage for the imposition of boundary conditions. If the reproducing conditions are satisfied and the shape functions in (7.42) possess the Kronecker-δ property, Dirichlet boundary conditions can be applied directly. The integration constraint determines area fractions for each particle at the boundary. These can be used to introduce Neumann boundary conditions into the system (Weißenfels 2019). Examples for the direct imposition of both boundary conditions can be found in Bode et al. (2020a, b) or Bode et al. (2021).

Remark Kilic and Madenci (2010) tries to justify the application of Neumann boundary conditions over a layer thickness. Since force densities are only determined internally in Peridynamics, force boundary conditions can only be applied as body forces. However, since the Peridynamic equations can be derived from the weak form, Neumann boundary conditions can be applied directly.

References

F. Bobaru, J.T. Foster, P.H. Geubelle, S.A. Silling, *Handbook of Peridynamic Modeling* (CRC press, 2016)

T. Bode, C. Weißenfels, P. Wriggers, Peridynamic Petrov-Galerkin method: a generalization of the peridynamic theory of correspondence materials. Comput. Methods Appl. Mech. Eng. **358**, 112636 (2020a)

T. Bode, C. Weißenfels, P. Wriggers, Mixed peridynamic formulations for compressible and incompressible finite deformations. Comput. Mech. **65**(5), 1365–1376 (2020b)

T. Bode, C. Weißenfels, P. Wriggers, A consistent peridynamic formulation for arbitrary particle distributions. Comput. Methods Appl. Mech. Eng. **374**, 113605 (2021)

M.S. Breitenfeld, P.H. Geubelle, O. Weckner, S.A. Silling, Non-ordinary state-based peridynamic analysis of stationary crack problems. Comput. Methods Appl. Mech. Eng. **272**, 233–250 (2014)

T. Breitzman, K. Dayal, Bond-level deformation gradients and energy averaging in Peridynamics. J. Mech. Phys. Solids **110**, 192–204 (2018)

S.R. Chowdhury, P. Roy, D. Roy, J.N. Reddy, A modified Peridynamics correspondence principle: removal of zero-energy deformation and other implications. Comput. Methods Appl. Mech. Eng. **346**, 530–549 (2019)

G.C. Ganzenmüller, S. Hiermaier, M. May, On the similarity of meshless discretizations of Peridynamics and smooth-particle hydrodynamics. Comput. Struct. **150**, 71–78 (2015)

M. Griebel, S. Knapek, G. Zumbusch, A. Caglar, *Numerische Simulation in Der Moleküldynamik: Numerik, Algorithmen, Parallelisierung* (Springer, Anwendungen, 2013)

P. Hartmann, C. Weißenfels, P. Wriggers, Application of enhanced peridynamic correspondence formulation for three-dimensional simulations at large strains, in *Virtual Design and Validation* (Springer, 2020), pp. 81–104

B. Kilic, E. Madenci, An adaptive dynamic relaxation method for quasi-static simulations using the peridynamic theory. Theor. Appl. Fract. Mech. **53**(3), 194–204 (2010)

D.J. Littlewood, Simulation of dynamic fracture using Peridynamics, finite element modeling, and contact, in *Proceedings of the ASME 2010 International Mechanical Engineering Congress and Exposition (IMECE)* (2010)

D.J. Littlewood, *Roadmap for Peridynamic Software Implementation* (SAND Report, Sandia National Laboratories, Albuquerque, NM and Livermore, CA, 2015)

R.W. Macek, S.A. Silling, Peridynamics via finite element analysis. Finite Elem. Anal. Des. **43**(15), 1169–1178 (2007)

E. Madenci, M. Dorduncu, A. Barut, N. Phan, Weak form of Peridynamics for nonlocal essential and natural boundary conditions. Comput. Methods Appl. Mech. Eng. **337**, 598–631 (2018)

H. Ren, X. Zhuang, Y. Cai, T. Rabczuk, Dual-horizon Peridynamics. Int. J. Numer. Methods Eng. **108**(12), 1451–1476 (2016)

S. Silling, D. Littlewood, P. Seleson, Variable horizon in a peridynamic medium. J. Mech. Mater. Struct. **10**(5), 591–612 (2015)

S.A. Silling, Reformulation of elasticity theory for discontinuities and long-range forces. J. Mech. Phys. Solids **48**(1), 175–209 (2000)

S.A. Silling, Stability of peridynamic correspondence material models and their particle discretizations. Comput. Methods Appl. Mech. Eng. **322**, 42–57 (2017)

S.A. Silling, A. Askari, Practical Peridynamics. Technical report, Sandia National Lab.(SNL-NM), Albuquerque, NM (United States) (2014)

S.A. Silling, R.B. Lehoucq, Peridynamic theory of solid mechanics. Adv. Appl. Mech. **44**, 73–168 (2010)

S.A. Silling, M. Epton, O. Weckner, J. Xu, E. Askari, Peridynamic states and constitutive modeling. J. Elast. **88**(2), 151–184 (2007)

Q. Tu, S. Li, An updated Lagrangian particle hydrodynamics (ULPH) for Newtonian fluids. J. Comput. Phys. **348**, 493–513 (2017)

M.R. Tupek, R. Radovitzky, An extended constitutive correspondence formulation of Peridynamics based on nonlinear bond-strain measures. J. Mech. Phys. Solids **65**, 82–92 (2014)

C. Weißenfels, Direct nodal imposition of surface loads using the divergence theorem. Finite Elem. Anal. Des. **165**, 31–40 (2019)

C.T. Wu, Kinematic constraints in the state-based Peridynamics with mixed local/nonlocal gradient approximations. Comput. Mech. **54**(5), 1255–1267 (2014)

C.T. Wu, C.K. Park, J.S. Chen, A generalized approximation for the meshfree analysis of solids. Int. J. Numer. Methods Eng. **85**(6), 693–722 (2011)

Chapter 8
Modeling Selective Laser Melting

Selective Laser Melting (SLM) is a laser powder bed fusion process that frequently comes into use for printing metal or ceramic structures. A positive connection between the layers can only be achieved by a remelt of the particles with the underlying layer. The key factor for successful printing is the control of the melt pool, which determines the shape and properties of the printed part. The parameters of the 3D printer must be set so that unwanted side effects can be excluded. The transition from the powder to the final printed part is a sequence of numerous physical processes that must be represented by the simulation. Suitable experimental tests are required to validate the models. Both aspects are presented in the first two sections. The differential equations are solved using SPH and the OTM method. The results of both schemes are compared and evaluated. The presented modeling approaches can also be used for Laser Metal Deposition or other laser powder bed fusion processes such as Electron Beam Melting or Selective Laser Sintering.

8.1 Physical Phenomena

In Selective Laser Melting (SLM), the laser passes its energy onto the powder particles via radiation. Parts of this energy are absorbed by the material, and some are returned to the environment by reflection. The energy or heat absorbed by the material is transported by conduction, convection, and radiation. Due to the large laser power, the temperature at the surface is very high and strong temperature gradients occur. If the temperature exceeds a limit value, the aggregate state changes and the material becomes liquid. At the uppermost layer, parts of the matter vaporize due to the very high temperatures. The interaction of surface tension, Marangoni convection, viscosity, and recoil pressure forms the melt and determines the geometry of a layer (Fig. 8.1). The 3D printing process causes changes on the microscale. The phase

C. Weißenfels, *Simulation of Additive Manufacturing using Meshfree Methods*, Lecture Notes in Applied and Computational Mechanics 97,
https://doi.org/10.1007/978-3-030-87337-0_8

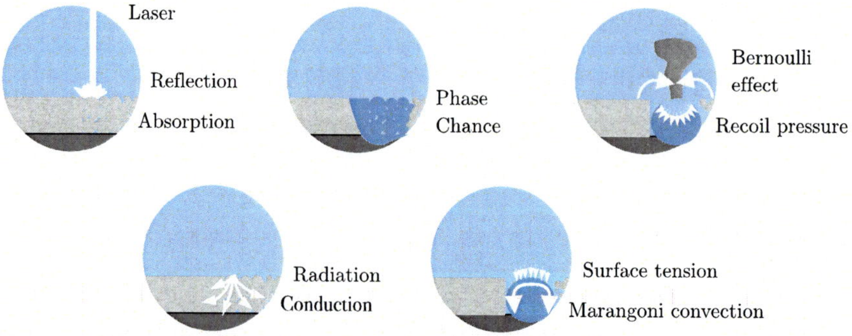

Fig. 8.1 Main physical phenomena during Selective Laser Melting

change leads to a reformation of the grain boundaries which depends on the cooling rate (Steen and Mazumder 2010). Their composition also influences the properties of the printed part. However, microscopic effects are not discussed. Information to this is given in Gu et al. (2012) and the references therein. Descriptions of individual physical effects can also be found in Yadroitsev (2009), King et al. (2015), Hebert (2016) or Yap et al. (2015).

8.1.1 Laser Matter Interaction

In Selective Laser Melting, the material is heated by the laser. The laser corresponds to electromagnetic radiation that propagates in a given direction. Metals consist of electrons that move around an ion lattice. The radiation excites the electrons to vibrate. This effect causes either the electrons themselves to radiate or the lattice structure to vibrate. The former leads to reflection or transmission, depending on the direction of the electron radiation. The latter corresponds to the absorbed energy. The absorption takes place over a certain depth and can be approximated by an intensity distribution I within the material. An estimation is provided for instance by the Beer-Lambert formula

$$I = I_0 \exp(-\beta z). \tag{8.1}$$

Therein, I_0 is the intensity at the surface and z the direction of radiation. The absorption coefficient β depends on the material, the wavelength of the laser, the radiation intensity, the temperature, the roughness of the surface, the angle of incidence of the radiation, and surface films (Steen and Mazumder 2010). This parameter is also called the extinction coefficient. The reflection of metallic materials is very large and the radiation does not penetrate deep into the material (Yadroitsev 2009). Materials showing such behavior are called opaque. In a powder bed, the reflected radiation hits other particles. Therefore, the total absorbed energy of a powder bed is larger compared to a pure metal block. The absorption also depends on the aggregate state. In the liquid phase, the value increases by a factor of 1.5 to 2 for metals in the infrared range (Prokhorov 2018).

8.1.2 Heat Transfer

The absorbed energy causes the lattice structure to vibrate. During thermal conduction, this energy is passed on to the adjacent lattice structures (Steen and Mazumder 2010) or released into the environment via conduction and radiation. Since a powder bed consists of a collection of particles, different forms of heat transport occur, see Fischer et al. (2003). These consist of conduction and radiation between materials and over the surrounding gas. The conduction via contact is assumed to be the main effect of heat transport (Ganeriwala and Zohdi 2016).

Thermal conductivity is a function of temperature and depends on the aggregate state. In the liquid phase, heat transport by convection occurs additionally due to the movement in the melt pool. An introduction to the basics of heat transfer can be found e.g. in Polifke (2009).

8.1.3 Heat Capacity and Phase Change

The energy required to vibrate the lattice structure can be related to temperature. The relationship depends on the material and is characterized by the heat capacity. This quantity describes the ratio of heat input $\triangle Q$ to temperature change $\triangle\Theta$. The division by mass defines the specific heat capacity c

$$c = \frac{\triangle Q}{m \triangle\Theta}. \tag{8.2}$$

The continuum mechanical implementation can be found in Sect. 3.3.3. If the vibrations in the lattice structures are very strong, collisions between the atoms occur. The lattice structures can be destroyed and the aggregate state of the material changes (Jiang and Wen 2011). The energy needed to dissolve the lattice structures corresponds to the latent heat and is defined as the heat input per mass required to change the aggregate state

$$L = \frac{\triangle Q}{m}. \tag{8.3}$$

8.1.4 Surface Tension and Wetting

Molecules embedded in their natural environment have the lowest energy. At the surface, molecules that are in contact with the molecules of the surrounding medium are in a state of higher energy. To minimize the total energy, the material tries to reduce contact with foreign molecules. This is only possible if the surface is minimal. The energy needed to increase an area by one is equal to the surface tension γ. The size of γ depends on temperature and surfactants. The gravitational force counteracts the surface tension. The dimensionless Bond number Bo compares these two effects. Additionally, a characteristic length l_{ca} can be determined (Bormashenko 2017). If this length is larger than the radius of the droplet R, the influence of surface tension dominates and the gravitational forces can be neglected

$$Bo = \frac{\rho g l^2}{\gamma}, \qquad R < l_{ca} = \sqrt{\frac{\gamma}{\rho g}}. \tag{8.4}$$

The Bond number depends on the density ρ, the gravitational acceleration g, the surface tension γ, and a characteristic length L which can be the radius.

Wetting corresponds to the behavior of liquids when they are in contact with a solid surface. A distinction is made between two cases. Either the liquid spreads completely or it forms a drop shape. When in contact with an ideal surface, the resulting behavior can be estimated from the spreading parameter Ψ

$$\Psi = \gamma_{SG} - (\gamma_{SL} + \gamma)\,. \tag{8.5}$$

An ideal surface is flat, chemically homogeneous, isotropic, insoluble, non-reactive and non-expandable (Bormashenko 2017). The spreading relates the surface tension between solid and gas γ_{SG} to the surface tension between liquid and gas γ and between solid and liquid γ_{SL}. Complete wetting occurs when Ψ is less than zero. In the other case, the surface is only partially wetted (Fig. 8.2). The contact angle between solid and liquid can be determined from the Young equation (Young 1805)

$$\cos\varphi = \frac{\gamma_{SG} - \gamma_{SL}}{\gamma}, \qquad \Psi = \gamma\,(\cos\varphi - 1)\,. \tag{8.6}$$

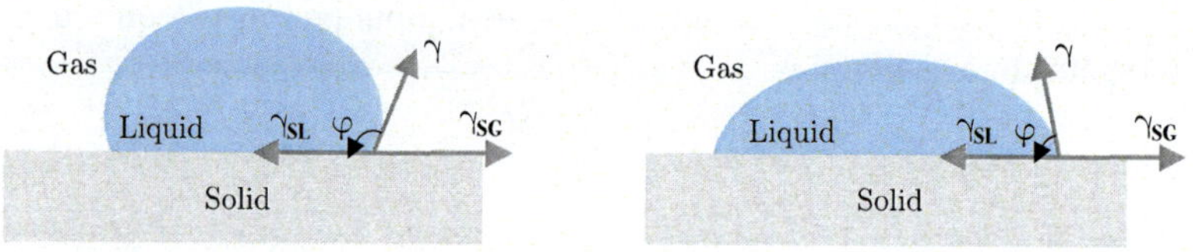

Fig. 8.2 (Left) Low wettability: $\Psi > 0$, contact angle between $90° \leq \varphi \leq 180°$. (Right) High wettability: $\Psi < 0$, contact angle $0° < \varphi < 90°$

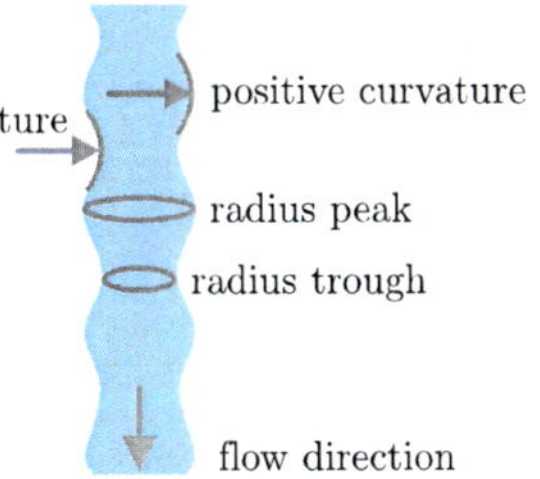

Fig. 8.3 Curvatures of a perturbed water jet

Microscopically, the cohesive forces of the molecules in the liquid want to form a droplet. The adhesive forces between the molecules of the solid and the molecules of the liquid cause the liquid to spread on the surface. The interactions are described via van der Waals forces. These can be divided into polar (Keesam forces and Debye forces) and disperse (London forces) ($\gamma = \gamma^p + \gamma^d$). Only polar and disperse fractions can interact with each other. Further information on surface tension and wetting can be found in Bormashenko (2017) or Adamson and Gast (1997).

Remark. A frequently occurring phenomenon is the breakage of a liquid jet into individual droplets. This effect is also known as Plateau Rayleigh instability. No water jet is perfectly cylindrical. The imperfections lead to a wave-like propagation in the axial direction, which can be approximated by the sum of individual sinusoidal functions. The surface tension is proportional to the curvature and hence inversely proportional to the radius. In the radial direction, the radius in the troughs is smaller than in the peaks. Thus, the surface tension in the trough is larger and pushes the liquid into the peaks. In the axial direction, the curvature in the troughs is negative and positive in the peaks due to the wave-like propagation. In this case, the surface tension pushes the liquid from the peaks into the troughs (Fig. 8.3). Thus, two different pressures compete during fluid flow. When the surface tension in the radial direction dominates, the fluid jet breaks up into droplets. The jet is assumed to be stable when the circumference computed from the diameter of the jet D is larger than the wavelength l in the axial direction

$$\frac{\pi D}{L} > 1. \tag{8.7}$$

A derivation of this criterion can be found e.g. in Chandrasekhar (2013).

8.1.5 Marangoni Convection and Viscosity

The laser only irradiates parts of the melt. Hence, the mass points at the surface have different temperatures. Since the surface tension is a function of temperature, a tangential force results from its gradient and leads to the Marangoni convection. The fluid particles flow from smaller to higher surface tension. For metallic materials, the

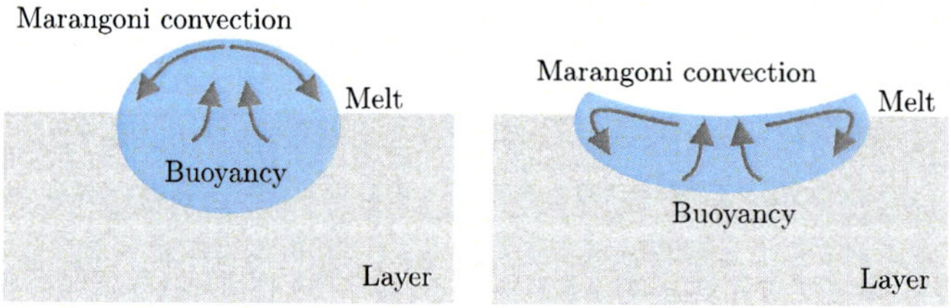

Fig. 8.4 Marangoni convection and buoyancy. (Left) Small Peclet and small Prandtl number. (Right) Small Peclet and high Prandtl number

coefficient γ decreases with temperature. Thus, the material flows from the laser to the outside (Fig. 8.4). In addition to the temperature, the concentration of dissolved substances influences the surface tension. With sulfur, for example, the coefficient γ increases with temperature.

Temperature also influences density. Density changes result in a thermal buoyancy-driven flow which points upwards (Fig. 8.4). In the Navier Stokes equations, this effect can be included by the Boussinesq approximation (e.g. Kleinstreuer 1997). The consequences from buoyancy are very small in Selective Laser Melting processes (Yadroitsev 2009).

The mass transport in the melt is influenced by the viscosity. The dynamic viscosity η determines the relation between an applied force F and the velocity of the flow v. The size of η can be specified by means of a viscous liquid between two plates (area A and distance h)

$$\eta = \frac{F\,h}{A\,v}. \tag{8.8}$$

The viscosity depends on temperature. This relation is usually assumed to be exponential (e.g. Limmaneevichitr and Kou 2000).

The geometry of the melt pool can be estimated using two dimensionless numbers. The Peclet number Pe describes the ratio of convection to conductivity

$$Pe = \frac{l\,v}{\alpha} = \frac{l\,v\,\rho\,c}{k}, \qquad \alpha = \frac{k}{\rho\,c}. \tag{8.9}$$

This value is calculated from the characteristic length l, the velocity v, and the thermal diffusivity α. The latter can also be related to the thermal conductivity k and the specific heat capacity c. If the Peclet number is very small, heat conduction dominates and the lower boundary of the melt pool has a concave shape (Yadroitsev 2009). As the Peclet number increases, the geometry at the bottom changes from a flat to a convex shape for very large values. The Prandtl number Pr describes the ratio of kinematic viscosity ν to thermal diffusivity

$$Pr = \frac{\nu}{\alpha} = \frac{\eta\,c}{k}, \quad \nu = \frac{\eta}{\rho}. \tag{8.10}$$

With small Prandtl numbers, heat conduction dominates and the melt pool forms spherically (8.4). At higher values, the Marangoni convection dominates and results in a wide and flat melt pool (Yadroitsev 2009).

8.1.6 Recoil Pressure

If the temperature exceeds a critical value, the upper layer vaporizes. The gaseous matter is ejected at high speed in the opposite direction of the laser radiation. This results in a recoil pressure on the melt, which depends strongly on the temperature (Bäuerle 2013; von Allmen and Blatter 2013). Analytical models for estimating the recoil pressure during vaporization can be found in Anisimov and Khokhlov (1995) or Klassen et al. (2014). A simplified approach is based on the Clausius Clapeyron equation, e.g. Khairallah et al. (2016). Therein, the relation between recoil and ambient pressure is weighted by a factor of 0.54

$$p^{rec} = 0.54 p_a \exp\left(\frac{\Delta H_{vap}}{R} \left(\frac{1}{\Theta_B} - \frac{1}{\Theta} \right) \right). \tag{8.11}$$

Θ_B corresponds to the boiling temperature. Instead of the enthalpy of vaporization ΔH_{vap} and the gas constant R, the equation can also be related to the Boltzmann constant k_B (Khairallah et al. 2016). The conversion is given by the Avogadro constant N_A, i.e. $R = N_A k_B$. The recoil pressure flattens the melt pool and is therefore a desired effect in Selective Laser Melting (Kruth et al. 2004). At very high laser power, the recoil pressure provides a strong indentation below the laser which is called keyhole mode laser melting (e.g. Rai et al. 2007).

A lot of energy is required for the phase change from liquid to gas. The vaporized material contains part of the absorbed energy of the laser. These lead to a reduction of the energy in the melt pool, which accounts for up to 70–90% of the laser power (Semak and Matsunawa 1997). This energy loss results in a reduction of the maximum temperature in the melt pool.

Remark 1 The high exit velocity of the vaporized material leads to a pressure difference in the build envelope. This phenomenon is also known as the Bernoulli effect (Fig. 8.5).

Remark 2 If the laser power is further increased, plasma forms above the melt pool which interacts with the laser (Bäuerle 2013).

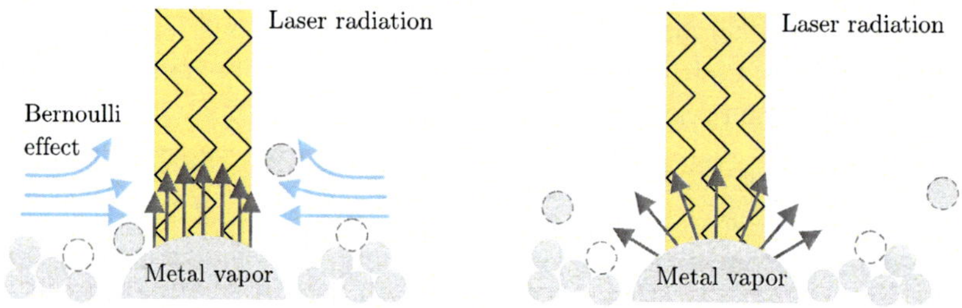

Fig. 8.5 Vaporization of material. (Left) Continuum flow. Low Knudsen number $\lambda < 1$. (Right) Molecular flow, $\lambda > 1$

8.1.7 Chemical Reactions

Another boundary layer effect that affects the quality of the printed part is oxidation or nitriding at the surface of the powder particles. The resulting ionic bonds hinder wetting. In addition, the melting temperature in the oxidized layer is significantly higher than in the pure metal. This can lead to brittle, porous parts in the printed component. For this reason, special gases such as argon or nitrogen are used in the build envelope to reduce the oxidation of the powder material (Yadroitsev 2009).

8.1.8 Powder Flow

During melting, the powder particles are set in motion. Controlling the flowability of the powder bed is, therefore, another factor in Selective Laser Melting that influences the quality of the printed part. The flowability depends on the contact forces between the particles and with the surrounding layers or walls. These forces mainly result from friction and cohesion. The forces depend largely on the shape, size, size distribution, and surface texture of the powder particles.

A measure for the particle flow is the Hausner factor (Hr) and the Carr index (Ci). Both measures are based on the ratio of tap (ρ_{td}) - to apparent density (ρ_{ad}).

$$Hr = \frac{\rho_{td}}{\rho_{ad}}, \qquad Ci = \frac{\rho_{td} - \rho_{ad}}{\rho_{ad}}. \tag{8.12}$$

A Hausner factor between 1.0 and 1.4 or a Carr index of 5–20 characterizes good flow characteristics. The flow property only describes the particle flow. For flowability, the environment in which the particles are located must also be taken into account. A spherical particle shape is advantageous for flowability. In addition, a more uniform powder bed can be generated (EPMA 2016).

The production of metallic powder can be realized in four different ways: chemically, by electrolysis, mechanically by grinding or crushing, or by atomization (Lawley 1978). Gas atomization (Dietrich et al. 2016) has proved to produce particles with very good properties for Selective Laser Melting (Yadroitsev 2009).

Remark. The formation of pores in the powder bed can lead to air pressure gradients (Prescott and Barnum 2000), which additionally influence the flowability.

8.2 Experimental Observations

In addition to specifying the geometry of the printed part and the material recipe, the process parameters need to be determined. In Selective Laser Melting, primarily the laser power, the scan rate, the diameter of the laser beam, the hatching, the height of the powder bed, and the sieve curve of the particles are to be determined. The right choice is critical to the quality of the printed part.

With SLM, negative side effects can occur which must be avoided during printing. Many of these effects can already be seen when printing a vector on a substrate (Figs. 8.6 and 8.7).

8.2.1 Denudation Zone

A very noticeable feature is the formation of a cleared area next to the remelted material. This area is called the denudation zone. The cleared area depends on the scan rate (Yadroitsev et al. 2010), the laser power, and the ambient pressure in the chamber (Matthews et al. 2016).

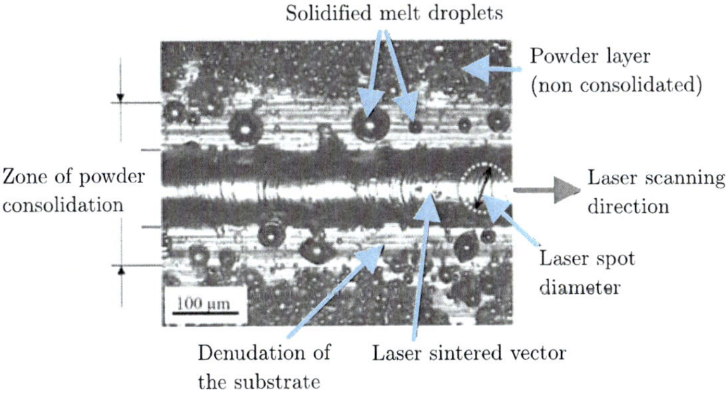

Fig. 8.6 Printing of a single vector on a substrate (Yadroitsev 2009). Clear formation of a denudation zone

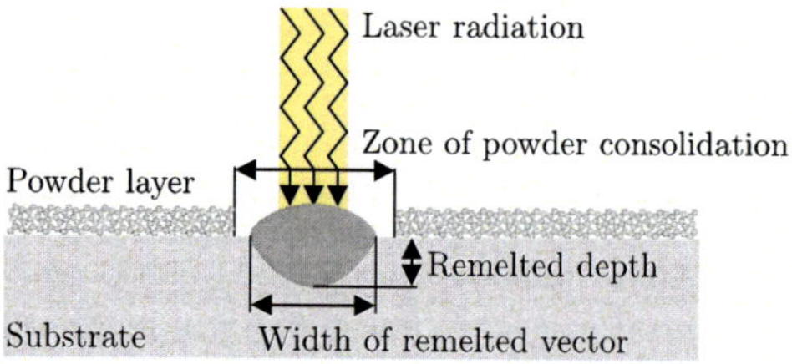

Fig. 8.7 Schematic description of a perfectly printed vector (cylindrical shape) on a substrate

According to Matthews et al. (2016) the vaporization of material at the top layer reasons the formation of the denudation zone. The shape of this zone depends on the Knudsen number Kn which describes the ratio of the mean free wavelength of the gas molecules λ to the characteristic length (here the diameter of a powder particle D)

$$Kn = \frac{\lambda}{D} = \frac{\eta}{p_0}\sqrt{\frac{\pi\, k_B\, \Theta}{2m}}. \tag{8.13}$$

This ratio can also be expressed by the dynamic viscosity η, the ambient pressure p_0, the mass of a gas atom in the chamber m, the temperature Θ, and the Boltzmann constant k_B. If the Knudsen number is less than 1, a continuum flow occurs. The vaporized particles flow vertically at high velocity out of the molten metal. The Bernoulli effect sucks in particles form the environment (Fig. 8.5). According to an estimate in Matthews et al. (2016) the flow of the rising jet can be turbulent. With a Knudsen number larger than 1, the molecular flow dominates. In this case, the vaporized metal particles escape more laterally from the melt and push the adjacent powder particles outwards (Fig. 8.5). In Bidare et al. (2018) it is shown that the direction of the outflowing metal vapor depends on the laser velocity. At a higher laser power, the metal particles are emitted to the rear and to the sides.

The denudation zone has a major impact on the process parameters. When printing the first layer on the substrate, the hatch distance should be larger than the denudation zone. This is the only way to provide enough powder particles for the next vector. If the distance is too small, the laser radiation primarily strikes the substrate or the already remelted neighboring vector. There, the reflection and the heat conduction are larger and less energy is introduced in the system (Yadroitsev and Smurov 2011). Printing the second layer, the entrainment of particles from the environment has the opposite influence. Due to the denudation zone of the previous layer, more particles are deposited in the current layer (Fig. 8.8). This leads to a strong increase in the remelted volume (Fig. 8.9).

8.2.2 *Irregular Tracks*

The goal of 3D printing using SLM is to generate continuous vectors of constant width that fuse with the previous layer. The cross-section view shows a vector possessing a

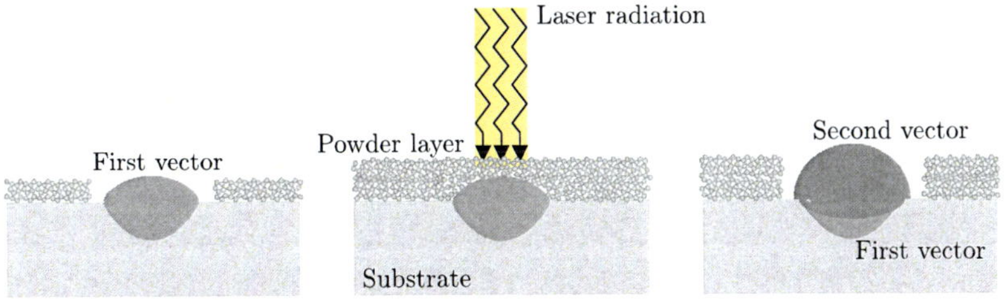

Fig. 8.8 Schematic description of printing a second vector

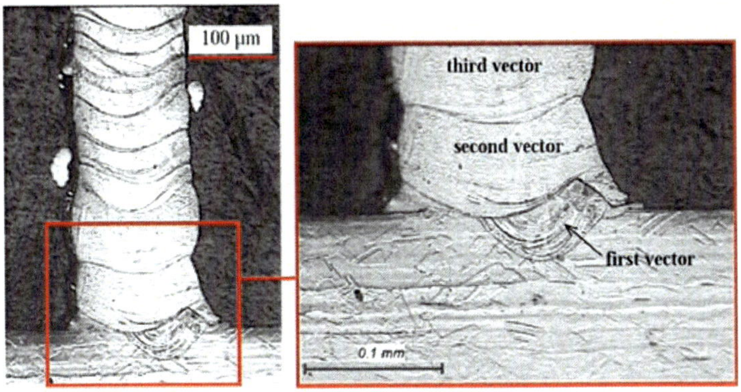

Fig. 8.9 Increase of volume due to printed vectors on top of the first vector (Yadroitsev 2009)

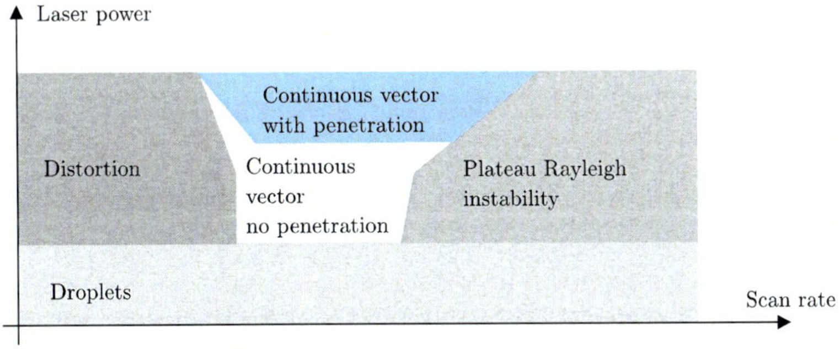

Fig. 8.10 Typical zones of forming continuous (blue, white) and non-continuous vectors (gray) in Selective Laser melting depending on the laser power and scan rate

cylindrical shape (Figs. 8.7 and 8.6). At certain process parameters, this cylindrical shape breaks up and agglomerates are formed, see e.g. Childs et al. (2005). This effect is called balling. A decisive criterion is the remelted depth. According to Yadroitsev et al. (2010) balling occurs either at a too low laser power P, a too high scan rate v, or a too-high energy e per running meter (Fig. 8.10). The latter size corresponds to the ratio of laser power to scan rate $e = P/v$.

If the laser power is too low, only the particles are melted, but there is no fusion with the underlying layer. The surface forces cause the vector either to take on an

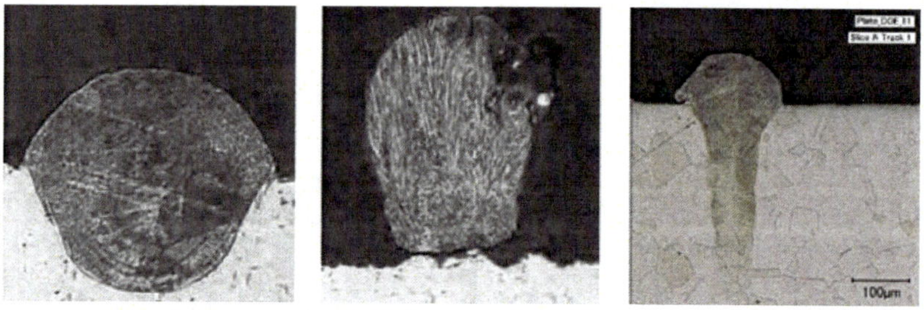

Fig. 8.11 (Left) Continuous cylindrical shape with penetration into the previous layer. (Middle) Irregular vector with no penetration into the previous layer (Yadroitsev 2009). (Right) Remelted vector at keyhole laser melting (King et al. 2014)

irregular shape or to break up into individual droplets. A temporal study on the formation of droplets shows different intermediate steps (Tolochko et al. 2004).

With sufficient laser power, the melting depth reduces when the scan rate increases (Fig. 8.11). In addition, the melt pool lengthens. Thus, the criterion from Eq. (8.7) can be violated and Plateau Rayleigh instabilities occur. In Yadroitsev et al. (2010) a criterion was developed that determines a critical size from the length and width of the melt pool and the remelted depth. This equation was experimentally confirmed in Gunenthiram et al. (2017).

If the energy input is very high, the melted area increases. Due to the high temperature, the viscosity of the melt decreases. The capillary forces are therefore very dominant and facilitate droplet formation (Yadroitsev et al. 2010). In addition, the recoil pressure increases, which renders the formation of irregular vectors more likely (Yadroitsev 2009).

8.2.3 *Pores and Surface Roughness*

Another undesired phenomenon in Selective Laser Melting is the development of pores in the material and the formation of rough surfaces. Pores influence the material properties and rough surfaces require post-treatment. Experimental investigations in Thijs et al. (2010) show that a small hatching distance and/or a low scan rate facilitate the formation of pores. If the laser power is too low, pores develop, because not all particles are melted and islands are formed at the surface (Bauereiß et al. 2014). The development of small pores in the interior of the material at very high laser power was reasoned by enclosed gas if the powder particles are gas atomized.

In Qiu et al. (2015) the formation of surface roughness is attributed to an oscillating behavior of the melt pool, which can possess a high frequency and amplitude. A dependence on the height of the powder layer was also determined. The reason for these instabilities is assigned to the interaction of Marangoni convection and recoil pressure. In addition, open and void-like pores were found in the evaluations of the surface texture.

The melt pool emits welding spatters which are directed forwards, upwards, or backward, measured in the direction of irradiation. The size of the spatter differs depending on the direction of ejection (Wang et al. 2017). Due to surface tension, spherical particles are formed during cooling. Their size exceeds the dimensions of the original particles in the powder bed by a multiple. If these particles are fused with the previous layer, cavities can form in the powder bed which cannot be filled during remelting (Wang et al. 2017).

At very high power, keyhole laser melting occurs, which results in a strong indentation below the laser. This indentation collapses after irradiation. This process also promotes the formation of pores (King et al. 2014).

8.2.4 *Residual Stresses*

The irradiation by the laser leads to the melting of the material. The prevented expansion of the melt creates compressive stress on the surrounding solid material. The yield stress decreases with increasing temperatures and plastic deformations occur more likely (Mercelis and Kruth 2006). During cooling, the prevented shrinkage generates tensile stresses (Gu et al. 2012) which again can exceed the yield stress. The tensile stresses are larger in the exposure direction than perpendicular to it (Gusarov et al. 2011). The residual stresses resulting from the prevented strains are independent of the scan rate (Shiomi et al. 2004) but depends on the hatch distance (Mercelis and Kruth 2006).

A general overview of the reasons and origins of residual stresses can be found, for example, in Withers and Bhadeshia (2001).

8.3 Laser Modeling

The laser radiates either in a pulsating or a continuous manner. In the former, the laser power at a point on the beam axis can be represented as a superposition of individual pulses (Fig. 8.12). With the definition of the maximum pulsed power P_{pulse}, the duration of radiation Δt_{pulse}, and the repetition frequency $f_{rep} = 1/\Delta t_{rep}$, the laser power can be represented as the sum of individual Heaviside functions H

$$P(t) = P_{pulse} \sum_{n=0}^{n_{pulse}} \left[H_n \left(t - \frac{n}{f_{rep}} \right) - H_n \left(t - \frac{n}{f_{rep}} - \Delta t_{pulse} \right) \right]. \tag{8.14}$$

The number of pulses is abbreviated with n_{pulse}. If the laser radiates continuously, the power is constant at all times. The continuous power P_{laser}, which energy is equivalent to the case of P_{pulse}, is determined by the maximum pulsed power, the repetition frequency, and the irradiation time

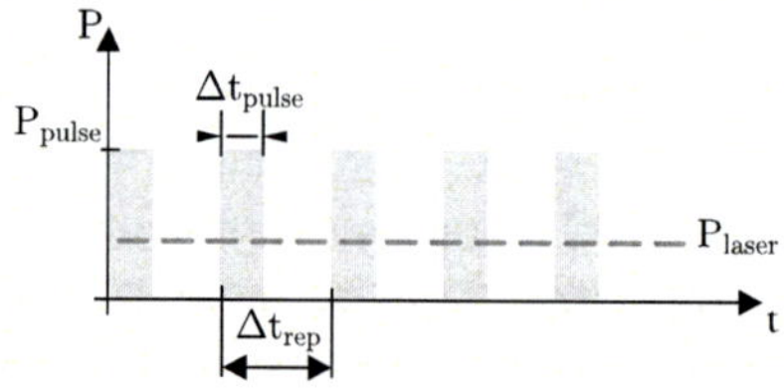

Fig. 8.12 Illustration of laser power over time for a laser in pulsed mode and energy equivalent laser power for a continuous mode

$$P_{laser} = f_{rep}\, \Delta t_{pulse}\, P_{pulse}. \tag{8.15}$$

The distribution of the power over the cross-section of the laser beam is generally not constant. The distribution in radial direction can be represented by an intensity function

$$P\,(t, r) = P\,(t)\; I_{rad}\,(r)\,. \tag{8.16}$$

This function depends on the radius r. The integral over the cross-section must result in one to distribute the whole laser power

$$\int_0^\infty I_{rad}\,(r)\;\mathrm{d}a = 1. \tag{8.17}$$

Many laser sources have a Gaussian distribution and the intensity function can be approximated depending on the radius of the laser beam R

$$I_{rad}\,(r) = \frac{2}{\pi R^2\,(1 - \exp(-2))} \exp\!\left(\frac{-2r^2}{R^2}\right). \tag{8.18}$$

A detailed introduction to laser technology and an overview of different types of lasers can be found in Hecht (2019).

The laser irradiation corresponds to a heat source that penetrates the material. For modeling, the irradiation can either be represented as electromagnetic field components using Maxwell equations (Zohdi 2012), or as discrete rays (Zohdi 2013). The radiation penetrates the material very quickly. The energy is only absorbed by a certain layer. Thus, irradiation can also be interpreted as an internal heat source. The penetration depth must be estimated. In the following two sections, the ray-tracing method and two modeling approaches of a volumetric heat source are presented.

Remark. In Zohdi (2014) four different approaches for modeling the laser irradiation are compared, which can also be interpreted as a sophistication. The simplest model estimates the temperature at a material point by the heat input. Influences from heat conduction are neglected. Thus, this approach is inappropriate for modeling Selective Laser Melting. The volumetric heat source and the ray-tracing method represent the next two accuracy levels. The most detailed modeling is based on the Maxwell equations.

8.3.1 Ray-Tracing

Ray-tracing is derived from the Maxwell equations (e.g. Zohdi 2013). The laser power is subdivided into discrete energy packets. Along the beam axis, the laser power is discretized in time. If the laser radiates continuously, the energy of a ray results from the multiplication of the power by a time increment Δt_r. To consider the intensity distribution of the laser in the cross-section, the laser power is weighted in the radial direction

$$E_r = \Delta t_r \, P_{laser} \, w_r \,(r_r)\,. \tag{8.19}$$

The weighting w_r corresponds to a spatial discretization and is based on the radial intensity function of the laser

$$w_r = \frac{I_{rad}\,(r_r)}{\sum_q^{n_{rpl}} I_{rad}\left(r_q\right)}. \tag{8.20}$$

This function is normalized, where n_{rpl} corresponds to the number of energy packages in radial direction. If the laser is modeled using energy packages, their propagation must also be determined. The current location of the ray $\mathbf{x}_r$ results from the integration in time of the propagation velocity of the laser $\mathbf{v}_r$, which in argon corresponds approximately to the speed of light

$$\dot{\mathbf{x}}_r = \mathbf{v}_r \approx 3 \cdot 10^8. \tag{8.21}$$

When the laser hits the material, only a part of the laser power is absorbed. The reflected contribution can be determined by the Fresnel equation

$$R_r = \frac{1}{2}\left[\left(\frac{\frac{\hat{n}^2}{\hat{\mu}}\cos\theta_r^{in} - a}{\frac{\hat{n}^2}{\hat{\mu}}\cos\theta_r^{in} + a}\right)^2 + \left(\frac{\cos\theta_r^{in} - \frac{1}{\hat{\mu}}a}{\cos\theta_r^{in} + \frac{1}{\hat{\mu}}a}\right)^2\right], \quad a = \left(\hat{n}^2 - \sin^2\theta_r^{in}\right)^{1/2}. \tag{8.22}$$

In the above equation, θ_r^{in} corresponds to the angle of incidence, $\hat{\mu}$ to the magnetic permeability, and $\hat{n}$ to the refractive index. Details of this equation can be found e.g. in Zohdi (2013).

8.3.2 Volumetric Heat Source

The laser radiation is absorbed by the uppermost layer, whereby the radiation decreases with the penetration depth. Thus, the laser power can also be interpreted as an internal heat source (Fig. 8.13). In this case, the intensity function must be extended by a component in the axial direction. This can, for instance, be realized independently of the radial direction

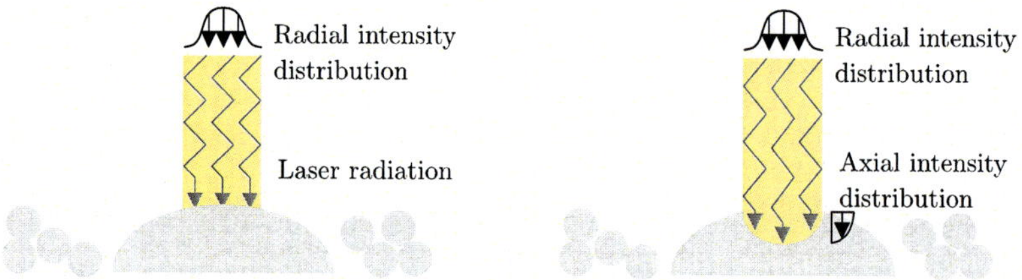

Fig. 8.13 (Left) Laser irradiation. (Right) Axial intensity distribution within the material using a volumetric heat source approach

$$P(t, r, z) = P(t)\; I_{rad}(r)\; I_{dep}(z)\,. \tag{8.23}$$

In order to divide the complete laser power, the integral over the whole volume must result in one

$$\int_0^L \int_0^\infty I_{rad}(r)\; I_{dep}(z)\;\mathrm{d}a\,\mathrm{d}z = 1. \tag{8.24}$$

The intensity function in axial direction can be formulated according to the Beer-Lambert formula (8.1),

$$I_{dep}(z) = \exp(-\beta z)\frac{\beta}{(1 - \exp(-\beta L))}. \tag{8.25}$$

This function is normalized to fulfill the condition from (8.24). The penetration depth L can be guessed by the size of the extinction coefficient β.

The powder bed possesses numerous cavities. In Gusarov et al. (2009), the coefficient β was therefore defined as a function of the porosity ϵ_b and the mean particle diameter D

$$\beta = \frac{2(1-\varepsilon_b)}{2\varepsilon_b\, D}. \tag{8.26}$$

A simulation approach for Selective Laser Melting based on this modified Beer-Lambert law can be found in Ganeriwala and Zohdi (2016). To consider the reflections in the powder bed, an analytical approach for the intensity distribution in the axial direction of the laser is developed in Gusarov et al. (2009)

$$I_{dep}(z) = -\beta\frac{\mathrm{d}q}{\mathrm{d}\xi\,(z)}. \tag{8.27}$$

The quantity q describes a dimensionless laser energy density and $\xi(z) = \beta z$ a dimensionless coordinate in axial direction. The exact specification of this intensity distribution can be found in Gusarov et al. (2009) or Wessels et al. (2018b). In addition, an alternative distribution of the laser radiation in the radial direction is proposed

$$I_{rad}(r) = \frac{3}{\pi R^2}\left(1-\frac{r}{R}\right)^2\left(1+\frac{r}{R}\right)^2. \tag{8.28}$$

Further approaches for volumetric heat sources can be found in Goldak et al. (1984). Therein, the models are specially developed for the simulation of laser welding.

8.4 Modeling of Material Response

The reactions of the material to the heat input can be described by the balance of momentum, the 1st law of thermodynamics, and suitable constitutive and kinematical equations. The melt can be assumed to be incompressible. The stress tensor is therefore split into a hydrostatic part p and an isochoric contribution $\boldsymbol{\sigma}^{iso}$

$$\rho\ddot{\mathbf{u}} = \rho\dot{\mathbf{v}} = -\,\mathrm{grad}\, p\mathbf{1} + \mathrm{div}\,\boldsymbol{\sigma}^{iso} + \rho\bar{\mathbf{b}}. \tag{8.29}$$

In addition, the influence of latent heat during phase change must be taken into account. For this, the formulation of the 1st law of thermodynamics in dependence on the enthalpy is preferred

$$\rho\dot{H} - D_{int} + \mathrm{div}\,\mathbf{q} - \rho r = 0. \tag{8.30}$$

8.4.1 Kinematics

The surface forces generate a fluid flow in the melt. Meshfree methods allow a consistent transition from solid deformations to fluid flow in the simulation. The material melts very quickly due to the high laser power and no large deformations are to be expected in the solid phase. Nevertheless, modeling based on finite deformations is suitable to ensure a consistent transition between the phases.

For metals, the temperature leads to a volumetric expansion of the material. This can be modeled by a split of the Jacobi determinant into a volumetric and isochoric part (Lu and Pister 1975)

$$\mathbf{F} = \mathbf{F}^{iso}\mathbf{F}^{vol}, \quad J = \det\mathbf{F}^{vol}, \quad J = J^e J^{th}, \quad J^{th} = \exp(3\alpha_\Theta\vartheta). \tag{8.31}$$

The parameter α_Θ corresponds to the coefficient of thermal expansion. In Selective Laser Melting, plastic deformations occur which can be regarded as isochoric. Hence, the isochoric deformation gradient is further subdivided into an elastic $\mathbf{F}^{iso,e}$ and a plastic $\mathbf{F}^{iso,p}$ contribution

$$\mathbf{F}^{iso} = \mathbf{F}^{iso,e}\mathbf{F}^{iso,p}. \tag{8.32}$$

During phase change the metallic bonds are dissolved (Jiang and Wen 2011). Since the volumetric stress results from the incompressibility, it can be assumed that only the isochoric strains are dissolved during phase change (Wessels et al. 2018b). Thus, $\mathbf{F}^{iso}$ can be set equal to the unit tensor $\mathbf{1}$ at phase transition which takes place at the limit temperature Θ_{phase}

$$\mathbf{F}^{iso} = \mathbf{1}, \qquad \text{if} \quad \Theta = \Theta_{phase}. \tag{8.33}$$

8.4.2 Constitutive Equations Solid Phase

Suitable thermo-elasto-plastic modeling is required to determine residual stresses in the material. The rheological model used in this approach is shown in Fig. 8.14. According to Goldak and Akhlaghi (2005) the range up to half the melting temperature Θ_{mel} can be described as rate-independent plasticity. At higher temperatures, the material behaves more and more viscous.

If the material is assumed to be weakly compressible in the solid phase, the pressure can be calculated from the bulk modulus K and the elastic Jacobi determinant J^e

$$p = -\frac{K(\Theta)}{J} \ln J^e, \quad \text{if} \quad \Theta < \Theta_{mel}. \tag{8.34}$$

The bulk modulus, like most material parameters, changes with temperature. Using a Neo-Hookian approach, the stress is calculated from the elastic isochoric right Cauchy Green Tensor $\mathbf{b}^{iso,e} = \mathbf{F}^{iso,e}\mathbf{F}^{iso,e\,\mathrm{T}}$

$$\boldsymbol{\sigma} = -p\mathbf{1} + \frac{\mu(\Theta)}{J}\left(\mathbf{b}^{iso,e} - \frac{1}{3}\operatorname{tr}\mathbf{b}^{iso,e}\mathbf{1}\right), \quad \text{if} \quad \Theta < \Theta_{mel}. \tag{8.35}$$

The elastic deformation gradient results from the evolution equation of the plastic flow

$$\mathbf{d}^p = \operatorname{sym}\left(\dot{\mathbf{F}}\mathbf{F}^{-1} - \dot{\mathbf{F}}^e\mathbf{F}^{e\,-1}\right) = \dot{\gamma}\frac{\partial f}{\partial \boldsymbol{\sigma}}. \tag{8.36}$$

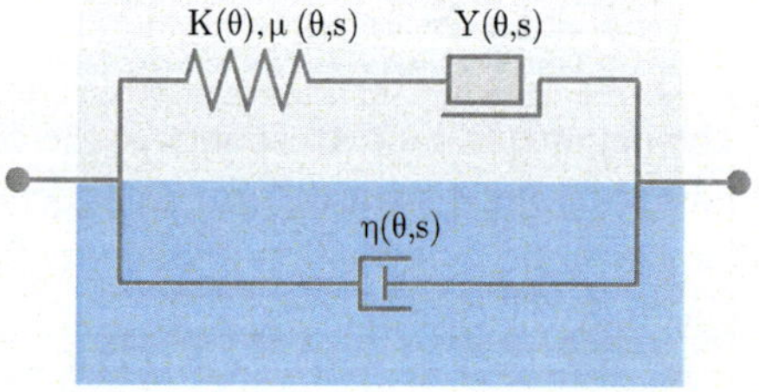

Fig. 8.14 Phase dependent rheological model. Solid phase (gray) and fluid phase (blue)

The elastic range is limited by the yield criterion. The yield stress Y and the hardening modulus H are assumed to decrease linearly with temperature (Wessels 2019)

$$f = \sqrt{\frac{3}{2}}\|\mathbf{s}\| - \left(Y\left[1 - \frac{\Theta - \Theta_0}{\Theta_{mel} - \Theta_0}\right] - H\left[1 - \frac{\Theta - \Theta_0}{\Theta_{mel} - \Theta_0}\right]\alpha\right) \leq 0. \tag{8.37}$$

For elasto-plasticity, the condition $\gamma \geq 0,\ f \leq 0$ with $\gamma\, f = 0$ must also be fulfilled. The rate of the isotropic hardening parameter α corresponds to the rate of the accumulated plastic strain γ, i.e. $\dot{\alpha} = \dot{\gamma}$.

In the range $\Theta < 0.5\,\Theta_{mel}$, γ is determined from the condition $f = 0$, if (8.37) is violated. In the transition zone $0.5\,\Theta_{mel} \leq \Theta < \Theta_{mel}$ the behavior is assumed to be viscoplastic and $\dot{\gamma}$ is calculated directly from the yield criterion

$$\dot{\gamma} = \frac{1}{2}\frac{f + |f|}{\mu}. \tag{8.38}$$

A detailed description of plasticity can be found in Lubliner (1990) or Simo (1998).

8.4.3 Constitutive Equations Fluid Phase

If the limit temperature Θ_{mel} is exceeded, the melt corresponds to a viscous liquid

$$\boldsymbol{\sigma} = -p\mathbf{1} + \frac{2\,\eta(\Theta)}{J}\left(\mathbf{d} - \frac{1}{3}\operatorname{tr}\mathbf{d}\,\mathbf{1}\right), \quad \text{if} \quad \Theta \geq \Theta_{mel}. \tag{8.39}$$

Since weakly incompressible behavior is assumed, only the isochoric part of the symmetric spatial velocity gradient $\mathbf{d}$ has to be considered. The parameter η corresponds to the dynamic viscosity of the melt. If the pressure is not calculated using the incompressibility condition, it can either be estimated using (8.34) or alternatively interpreted as a barotropic fluid. In the second case, the pressure is computed from (6.72)

$$p = \frac{\rho\, c_0^2}{7}\left[\left(\frac{\rho}{\rho_0}\right)^7 - 1\right], \quad \text{if} \quad \Theta \geq \Theta_{mel},$$

and the density is determined from the continuity equation (3.7)

$$\dot{\rho} + \rho \operatorname{div}\mathbf{v} = 0.$$

8.4.4 Thermal Constitutive Equations

In the 1st law of thermodynamics, the specific enthalpy and the heat conduction are required. Neglecting the Gough-Joule effect, a direct relationship between the specific enthalpy and the specific heat capacity c in the material results

$$H = c(\Theta)\dot{\vartheta}. \tag{8.40}$$

The contribution from latent heat can be easily integrated into the description of the specific heat capacity using the apparent heat approach (Bonacina et al. 1973)

$$c_{app} = \begin{cases} c_{s,f} & \text{if } \Theta < \Theta_s \\ 0.5(c_{s,f} + c_{f,g}) + L_{mel,vap}(\Theta_{liq,dew} - \Theta_{sol,bub})^{-1} & \text{if } \Theta_{sol,bub} \leq \Theta \leq \Theta_{liq,dew} \\ c_{f,g} & \text{if } \Theta > \Theta_{liq,dew} \end{cases}. \tag{8.41}$$

This form assumes a non-isothermal phase change. The state of the material does not change at a limit temperature but within a temperature range. The modeling requires an explicit specification of the solidus Θ_{sol} and liquidus Θ_{liq} temperature at the transition from solid to liquid and of the bubble Θ_{bub} and dew temperature Θ_{dew} in the case of the phase change from liquid to gaseous.

The heat conduction is assumed to be isotropic leading to a simplified model based on the thermal conductivity coefficient k

$$\mathbf{q} = -k(\Theta)\,\mathrm{grad}\,\vartheta. \tag{8.42}$$

Due to the high temperatures in the material, the contribution from dissipation due to plastic deformations is negligibly small. If the temperature at the surface exceeds a critical value, the loss of energy due to vaporized material leads to a cooling effect which can be approximated using the enthalpy of vaporization ΔH_{vap}, the molar mass M, the gas constant R, the sticking coefficient A, and the recoil pressure p^{rec} (Khairallah et al. 2016)

$$m\,r = -\frac{0.82\,\Delta H_{vap}\,A\,p^{rec}}{\sqrt{2\pi\,M\,R\,\Theta}}. \tag{8.43}$$

8.4.5 Neumann Boundary Conditions

The Neumann boundary conditions for the mechanical differential equation are composed of the surface tension, the Marangoni convection and the recoil pressure

$$\mathbf{t} = -\left[2\gamma(\Theta)\,\kappa + p^{rec}\,(\Theta)\right]\mathbf{n} + \mathrm{grad}_{\,S}\gamma(\Theta). \tag{8.44}$$

The vector $\mathbf{n}$ corresponds to the normalized normal vector. To recoil pressure p^{rec} computes from (8.11). The heat exchange over the surface is modeled using the thermal coefficient h

$$\bar{q} = -h\,(\Theta - \Theta_0) \tag{8.45}$$

If only a part of the whole underlying layers is considered, the coefficient must be a function of the missing layer l ($h := f(l)$). The influence of the radiation over the surface can be considered negligibly small (Khairallah et al. 2016).

8.5 Solution Using SPH

With Smoothed Particle Hydrodynamics, the simulation of fluid flow is usually based on a description with respect to the deformed configuration in combination with an explicit time integration scheme. The strong form is directly discretized. The quantities at the time step n are computed from the current neighborhoods which are redetermined at every step. To simplify the notation, an indication of the time step and the position vector is omitted, like for the volume $v_I := v_n(\mathbf{x}_{I\,n})$. If a specification is necessary it is indicated at the appropriate place.

8.5.1 *Spatial Discretization*

In the simulation using SPH, the individual particles in the solid phase are assumed to be rigid. If the critical temperature is exceeded, the primary variable $\mathbf{v}$ is calculated from (3.14)

$$-\operatorname{grad} p + 2\eta \operatorname{div} \operatorname{grad}^{sym} \mathbf{v} + \rho\bar{\mathbf{b}} - \rho\dot{\mathbf{v}} = 0 \qquad \Theta \geq \Theta_{mel}.$$

As already shown in Chap. 4, the discretized mechanical differential equation can be formulated in terms of a force balance evaluated at each particle

$$\mathbf{f}_I^s + \mathbf{f}_I^g + \mathbf{f}_I^t = \begin{cases} \mathbf{f}_I^n & \text{boundary particle} \\ 0 & \text{inner particle} \end{cases} . \tag{8.46}$$

In order to conserve global balance of momentum and angular momentum, the internal force due to the state of stress inside the material computes from (6.71) and (6.74)

$$\mathbf{f}_I^s = \sum_{J=1}^{N_I} \left[(p_J + p_I) - \eta \frac{\rho_I + \rho_J}{\rho} \frac{(\mathbf{v}_I - \mathbf{v}_J)\cdot(\mathbf{x}_I - \mathbf{x}_J)}{\|\mathbf{x}_I - \mathbf{x}_J\|^2 + \epsilon^2} \right] \frac{\partial W\,(\mathbf{x}_I - \mathbf{x}_J)}{\partial \mathbf{x}_I} v_I v_J . \tag{8.47}$$

The numeric parameter ϵ is set in relation to the smoothing length h, i.e. $\epsilon = 0.01\,h$ (Cleary and Monaghan 1999). The density is determined by the solution of the continuity equation (3.7). To ensure a constant density at constant velocity, the approach from (6.73) is chosen

$$\dot{\rho}_I = -\rho_I \sum_{J=1}^{N_I} (\mathbf{v}_J - \mathbf{v}_I) \cdot \frac{\partial W(\mathbf{x}_I - \mathbf{x}_J)}{\partial \mathbf{x}_I} v_J .$$

Assuming a barotropic fluid, the relationship between pressure and density at the particle I results from (6.72)

$$p_I = \frac{\rho_I\, c_0^2}{7} \left[\left(\frac{\rho_I}{\rho_0} \right)^7 - 1 \right] .$$

The forces from the dead load and the inertia term are determined only from quantities at the particle and corresponds to (4.16)

$$\mathbf{f}_I^t = m_I \ddot{\mathbf{u}}_I , \quad \mathbf{f}_I^g = -m_I \bar{\mathbf{b}}, \quad \text{with} \quad m_I = \rho_I v_I .$$

With SPH, the surface is not directly defined. However, the integration constraint can be employed to identify particles at the boundary (Weißenfels 2019). If this requirement is satisfied, only these particles have a non-zero resulting normal vector. Moreover, the norm of $\bar{\mathbf{n}}_I$ is equal to the surface fraction of the corresponding particle. Thus, surface loads can be applied directly. From (6.23) the resulting normal vector of a particle can be estimated

$$\bar{\mathbf{n}}_I = v_I \sum_{J=1}^{N_I} \frac{\partial W\,(\mathbf{x}_I - \mathbf{x}_J)}{\partial \mathbf{x}_J} v_J . \tag{8.48}$$

This approach is also used in Randles and Libersky (1996) or Hirschler et al. (2017) without mentioning the relationship to the integration constraint. In one-point methods, particles near the boundary often possess a resulting normal vector. Its value is smaller than at the boundary. Particles below a critical value are considered to be in the interior

$$\bar{\mathbf{n}}_I \equiv 0, \quad \text{if} \quad \|\bar{\mathbf{n}}_I\| < \frac{v_I}{5h} . \tag{8.49}$$

If the particles have a resulting normal vector unequal to zero, the forces from surface tension and Marangoni convection are imposed. If the temperature is higher than the boiling temperature, the influence of the recoil pressure is added

$$\mathbf{f}_I^n = -\left[2\gamma(\Theta_I)\,\bar{\kappa}_I + p^{rec}\,(\Theta_I) \right] \bar{\mathbf{n}}_I + \operatorname{grad}_S \gamma_I \,\|\bar{\mathbf{n}}_I\| . \tag{8.50}$$

The force resulting from surface tension is determined by the mean curvature

$$\kappa_I = \frac{1}{2}\left(\kappa_I^1 + \kappa_I^2\right). \tag{8.51}$$

The curvature, in turn, computes from the derivative of the normal vectors with respect to the local coordinates of the surface multiplied by the tangential base vectors $\mathbf{a}_{\beta\, I}$

$$B_{I\,\alpha\beta} = \frac{\partial \mathbf{n}_I}{\partial \xi^\alpha}\cdot \mathbf{a}_{\beta\, I}, \quad \mathbf{a}_{\beta\, I} = \frac{\partial \mathbf{x}_I}{\partial \xi^\beta}. \tag{8.52}$$

Local coordinates are not defined in SPH. Hence, an alternative way to determine the curvature is needed. In Fürstenau et al. (2019a), first a global curvature tensor is computed based on the corrected derivatives according to (6.43)

$$\mathbf{K}_I = \sum_{J=1}^{N_I}\left(\frac{\widetilde{\mathbf{n}}_J}{\|\widetilde{\mathbf{n}}_J\|} - \frac{\widetilde{\mathbf{n}}_I}{\|\widetilde{\mathbf{n}}_I\|}\right) \otimes \frac{\partial W(\mathbf{x}_I - \mathbf{x}_J)}{\partial \mathbf{x}_I}\mathbf{L}_I^{-1} v_J. \tag{8.53}$$

Therein, averaged normal vectors are employed

$$\widetilde{\mathbf{n}}_I = \frac{1}{W}\sum_{J=1}^{N_I} W\left(\mathbf{x}_I - \mathbf{x}_J\right)\bar{\mathbf{n}}_J, \quad W = \sum_{J=1}^{N_I} W\left(\mathbf{x}_I - \mathbf{x}_J\right). \tag{8.54}$$

Afterwards, the tensor $\mathbf{K}$ is transformed into a local coordinate system. The corresponding base vectors $\mathbf{g}_I$ can be constructed from the resulting normal vector and a cartesian base vector $\mathbf{e}_i$

$$\mathbf{g}_I = \left\{\widetilde{\mathbf{n}}_I, \widetilde{\mathbf{t}}_I^1, \widetilde{\mathbf{t}}_I^2\right\}, \quad \widetilde{\mathbf{t}}_I^1 = \frac{\widetilde{\mathbf{n}}_I}{\|\widetilde{\mathbf{n}}_I\|} \times \mathbf{e}_i, \quad \text{and} \quad \widetilde{\mathbf{t}}_I^2 = \frac{\widetilde{\mathbf{n}}_I}{\|\widetilde{\mathbf{n}}_I\|} \times \widetilde{\mathbf{t}}_I^1. \tag{8.55}$$

The matrix $\mathbf{T}$ transforms the components of $\mathbf{K}$ to the local curvature matrix. Its components result from the multiplication of the cartesian base vectors with there local counterparts $T_{I\,ij} = \mathbf{e}_i \cdot \mathbf{g}_I^j$

$$\mathbf{B}_I = \mathbf{T}_I^{\mathrm{T}}\mathbf{K}_I\mathbf{T}_I. \tag{8.56}$$

The local curvature tensor $\mathbf{B}_I$ can be divided into a normal and a tangential part. The eigenvalues of the tangential part correspond to the principal curvatures κ_I^1, κ_I^2 (Fig. 8.15). In order to better approximate the curvature, the principal curvatures are also averaged

$$\tilde{\kappa}_I = \frac{1}{W}\sum_{J=1}^{N_I} \kappa_J W\left(\mathbf{x}_I - \mathbf{x}_J\right). \tag{8.57}$$

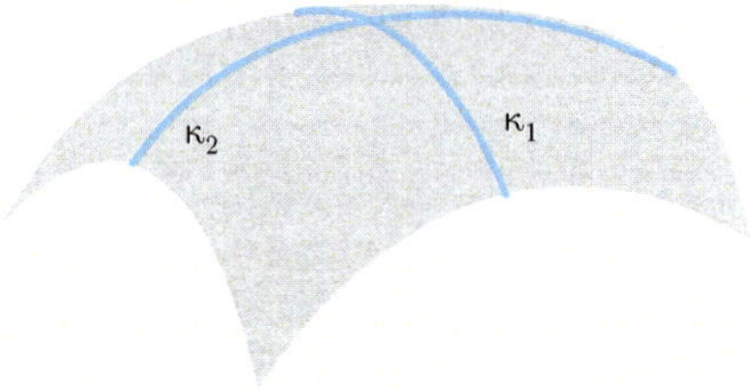

Fig. 8.15 Principle curvatures of a surface in space

The force due to Marangoni convection in (8.50) results from the derivative of the surface tension with respect to local tangential coordinates. The computation can be realized analogously to the curvature calculation. Starting from the derivative with respect to the global coordinates

$$\operatorname{grad} \gamma_I = \sum_{J=1}^{N_I} \gamma_J \frac{\partial W(\mathbf{x}_I - \mathbf{x}_J)}{\partial \mathbf{x}_I} \mathbf{L}_I^{-1} v_J \tag{8.58}$$

the normal vectors can be used to compute the tangential gradient

$$\operatorname{grad}^S \gamma_I = \operatorname{grad} \gamma_I - \left(\operatorname{grad} \gamma_I \cdot \frac{\widetilde{\mathbf{n}}_I}{\|\widetilde{\mathbf{n}}_I\|} \right) \frac{\widetilde{\mathbf{n}}_I}{\|\widetilde{\mathbf{n}}_I\|}. \tag{8.59}$$

To determine the temperature, the thermal differential equation (3.38) has to be solved

$$\rho \dot{H} - k \operatorname{div} \operatorname{grad} \vartheta - \rho r = 0.$$

To better approximate the influence of latent heat, the specific enthalpy is chosen as the solution function. The temperature is subsequently determined from an empirical relationship between specific enthalpy and temperature. In the discrete form, the differential equation can be formulated as the equilibrium of powers that have to be conserved at each particle separately. The discrete 1st law of thermodynamics postulates that the contribution of the specific enthalpy P_I^h together with the divergence of heat flow P_I^q and the inner heat source P_I^r must correspond to the power density at the Neumann boundary P_I^n

$$P_I^h + P_I^q + P_I^r = \begin{cases} P_I^n & \text{boundary particle} \\ 0 & \text{inner particle} \end{cases}. \tag{8.60}$$

Similar to (8.47) the contribution of the divergence of heat flow is alternatively discretized employing two first order derivatives (Cleary and Monaghan 1999)

$$P_I^q = -k \sum_{J=1}^{N_I} 2\,(\vartheta_J - \vartheta_I) \frac{\mathbf{x}_J - \mathbf{x}_I}{\|\mathbf{x}_I - \mathbf{x}_J\|^2 + \epsilon^2} \cdot \frac{\partial W(\mathbf{x}_I - \mathbf{x}_J)}{\partial \mathbf{x}_I} v_I v_J. \tag{8.61}$$

The power from the specific enthalpy results directly from the values defined at the particle

$$P_I^h = v_I \dot{H}_I. \tag{8.62}$$

Additionally, the cooling effect (8.43) is added to particles which are influenced by the recoil pressure

$$P_I^r = -\frac{0.82\, \Delta H_{vap}\, A\, p_I^{rec}}{\sqrt{2\, \pi\, M\, R\, \Theta_I}}. \tag{8.63}$$

The Neumann boundary condition corresponds to the laser entry. Here, a Beer-Lambert volumetric heat source from Sect. 8.3.2 is employed to assign a power to the particles below the laser

$$P_I^n = P_{laser} \frac{2}{\pi R^2} \frac{\beta}{1 - \exp(-\beta z)} \exp(-\beta\, z) \exp\left(-2\frac{r^2}{R^2}\right). \tag{8.64}$$

8.5.2 Temporal Discretization

The discretized equations contain time derivatives. To compute the velocities and the position vectors in the next time step, the Verlet algorithm is employed

$$\mathbf{v}_{I\,n+1} = \mathbf{v}_{I\,n-1} + 2\Delta t \dot{\mathbf{v}}_{I\,n}, \qquad \mathbf{x}_{I\,n+1} = \mathbf{x}_{I\,n} + \Delta t \mathbf{v}_{I\,n} + \frac{1}{2}(\Delta t)^2 \dot{\mathbf{v}}_{I\,n}. \tag{8.65}$$

The densities at the next time step are calculated analogously

$$\rho_{I\,n+1} = \rho_{I\,n-1} + 2\Delta t \dot{\rho}_{I\,n}. \tag{8.66}$$

To prevent disturbed solutions, after certain time steps the quantities are updated using the forward Euler time integration scheme (Crespo et al. 2015)

$$\mathbf{v}_{I\,n+1} = \mathbf{v}_{I\,n} + \Delta t \dot{\mathbf{v}}_{I\,n}, \qquad \mathbf{x}_{I\,n+1} = \mathbf{x}_{I\,n} + \Delta t \mathbf{v}_{I\,n} + \frac{1}{2}(\Delta t)^2 \dot{\mathbf{v}}_{I\,n}. \tag{8.67}$$

This concept is also employed for the density update at these time steps

$$\rho_{I\,n+1} = \rho_{I\,n} + \Delta t \dot{\rho}_{I\,n}. \tag{8.68}$$

The specific enthalpy in the next time step is determined only by the forward Euler time integration scheme

$$H_{I\,n+1} = H_{I\,n} + \Delta t \dot{H}_{I\,n}. \tag{8.69}$$

In explicit integration schemes, the time step size must be limited to be numerically stable (Courant et al. 1928; Hughes 1987 or Belytschko et al. 1976). Each contribution provides a maximum time step size. Four different cases must be considered when modeling Selective Laser Melting. These result from the contribution of the maximum acceleration within a time step Δt_a as well as the viscosity Δt_v in (8.47), see also (Monaghan and Kos 1999)

$$\Delta t_a = \sqrt{\frac{h}{\ddot{\mathbf{u}}_{max}}}, \quad \Delta t_v = \frac{h}{c_0 + \max\left(\frac{h(\mathbf{v}_I - \mathbf{v}_J)\cdot(\mathbf{x}_I - \mathbf{x}_J)}{\|\mathbf{x}_I - \mathbf{x}_J\|^2 + \eta^2}\right)}, \tag{8.70}$$

the surface tension Δt_γ and from the thermal equation Δt_Θ (Fürstenau et al. 2019b)

$$\Delta t_\gamma = \sqrt{\frac{h^3 \rho_0}{2\pi\, \gamma_{max}}}, \quad \Delta t_\Theta = \frac{h^2 \rho_0\, c}{2\, k}. \tag{8.71}$$

The smallest value from the four contributions determines the maximum permissible time step size

$$\Delta t = 0.25 \min\left(\Delta t_a, \Delta t_v, \Delta t_\gamma, \Delta t_\Theta\right). \tag{8.72}$$

8.6 Solution Using OTM

The solution using the Optimal Transportation Meshfree method is based on an updated Lagrangian description together with an implicit time integration scheme. To ease the notation, an indication of the time step and the position vector is omitted, like

$$N_{I,p} := N_I(\mathbf{x}_{pn}), \quad \frac{\partial N_{I,p}}{\partial \mathbf{x}} := \frac{\partial N_I(\mathbf{x}_{pn})}{\partial \mathbf{x}_{pn+1}}, \quad v_p = v_{pn+1}(\mathbf{x}_{pn}). \tag{8.73}$$

If the exact time step needs to be specified, it is reintroduced at the appropriate place.

8.6.1 Spatial Discretization

The OTM algorithm is based on the weak form. The discretization of the mechanical differential equation (3.21)

$$\int_{B_t} \left[\rho\, \boldsymbol{\eta} \cdot (\ddot{\mathbf{u}} - \bar{\mathbf{b}}) + \operatorname{grad} \boldsymbol{\eta} \cdot \boldsymbol{\sigma}\right] \mathrm{d}v = \int_{\partial B_t} \boldsymbol{\eta} \cdot \bar{\mathbf{t}}\, \mathrm{d}a.$$

can be specified in the final form as the equilibrium of nodal forces, see also (4.2) and (4.3)

$$\mathbf{f}_I^s + \mathbf{f}_I^g + \mathbf{f}_I^t = \begin{cases} \mathbf{f}_I^n & \text{boundary node} \\ 0 & \text{inner node} \end{cases} . \tag{8.74}$$

The internal force due to the state of stress in the material $\mathbf{f}_I^s$, the dead load $\mathbf{f}_I^g$, and the inertia term $\mathbf{f}_I^t$

$$\mathbf{f}_I^\sigma = \sum_{p=1}^{N_I^{inf}} \sigma_p \frac{\partial N_{I,p}}{\partial \mathbf{x}} v_p, \quad \mathbf{f}_I^g = -\sum_{p=1}^{N_I^{inf}} N_{I,p} \bar{\mathbf{b}} m_p, \quad \mathbf{f}_I^t = \sum_{p=1}^{N_I^{inf}} N_{I,p} \ddot{\mathbf{u}}_p m_p \tag{8.75}$$

must correspond to the force due to the Neumann boundary condition $\mathbf{f}_I^n$ at each node. The conservation of mass is automatically fulfilled in the OTM method, since $m_p = m_{p\,n+1} = const$. Depending on the phase, the stress is either calculated from (8.35) or (8.39). The solution scheme to determine the stress in case of plasticity or visco-plasticity is based on the algorithm given in Korelc and Stupkiewicz (2014). The implementation can also be found in Wessels (2019).

The imposition of boundary conditions in meshfree methods always requires a special treatment. As mentioned in the previous section, boundary nodes can be identified using the integration constraint. If this criterium is fulfilled, only a node at the boundary possesses a resulting normal vector. Its norm corresponds to the area of that node and the sum of all nodal areas results in the total surface of the body (Weißenfels 2019)

$$\bar{\mathbf{n}}_I = \sum_{p=1}^{N_I^{inf}} \frac{\partial N_{I,p}}{\partial \mathbf{x}} v_p, \quad a_I = \|\bar{\mathbf{n}}_I\|. \tag{8.76}$$

To avoid an integration over the surface, the forces at the boundary can alternatively be calculated from a volume integral using the divergence theorem, like for the contribution from the surface tension

$$\int_{\partial B_t^\sigma} \boldsymbol{\eta} \cdot \gamma \kappa \mathbf{n} \; da = \int_{B_t} \text{div} \; (\gamma \kappa \boldsymbol{\eta}) \; dv. \tag{8.77}$$

If the surface tension γ is assumed to be constant and if the change of the curvature is neglected, the force results from the derivative of the shape function

$$\mathbf{f}_I^{sur} = -\sum_{p=1}^{N_I^{inf}} 2\, \gamma \tilde{\kappa}_p \frac{\partial N_{I,p}}{\partial \mathbf{x}} v_p. \tag{8.78}$$

The contribution from the recoil pressure can be calculated analogously

$$\mathbf{f}_I^{rec} = -\sum_{p=1}^{N_I^{inf}} p_p^{rec} \frac{\partial N_{I,p}}{\partial \mathbf{x}} v_p. \tag{8.79}$$

The tangential force due to Marangoni convection corresponds to the derivative of the surface tension coefficient with respect to local surface coordinates. Local coordinates are usually not available in meshfree methods. Therefore, as in the previous section, the derivative with respect to the global coordinates is determined first. The tangential force computes from the projection on the tangential plane with the help of the normal vector at the corresponding node

$$\mathbf{f}_I^{mar} = \|\bar{\mathbf{n}}_I\| \sum_{p=1}^{N_I^{inf}} \operatorname{grad}\gamma\,|_p \left(\mathbf{1} - \frac{\bar{\mathbf{n}}_I}{\|\bar{\mathbf{n}}_I\|} \otimes \frac{\bar{\mathbf{n}}_I}{\|\bar{\mathbf{n}}_I\|}\right), \quad \operatorname{grad}\gamma\,|_p = \sum_{I=1}^{N_p^{sup}} \frac{\partial N_{I,p}}{\partial \mathbf{x}} \gamma_I. \tag{8.80}$$

The sum of all contributions corresponds to the force due to Neumann boundary conditions $\mathbf{f}_I^n = \mathbf{f}_I^{sur} + \mathbf{f}_I^{mar} + \mathbf{f}_I^{rec}$.

The surface tension requires the mean curvature κ. Analogous to SPH, κ is determined from the global curvature tensor $\mathbf{K}$

$$\mathbf{K}_p = \sum_{I=1}^{N_p^{sup}} \frac{\bar{\mathbf{n}}_I}{\|\bar{\mathbf{n}}_I\|} \otimes \frac{\partial N_{I,p}}{\partial \mathbf{x}}. \tag{8.81}$$

Using the resulting normal vector (8.76) a local, orthonormal coordinate system can be defined. These base vectors transform $\mathbf{K}$ to the local curvature tensor $\mathbf{B}$. The eigenvalues of the tangential part of $\mathbf{B}$ correspond to the principal curvatures. Also with this approach, an averaging of the curvatures over the neighboring integration points N_p^{ngh} proved to be advantageous (Wessels et al. 2018b)

$$\tilde{\kappa}_p = \frac{1}{N_p^{ngh}} \sum_{q=1}^{N_p^{ngh}} \kappa_q. \tag{8.82}$$

In the thermal differential equation (3.41)

$$\int_{B_t} \left(\tau[\rho\dot{H} - \rho r] + k \operatorname{grad}\tau \cdot \operatorname{grad}\vartheta\right) \mathrm{d}v + \int_{\partial B_t^{N_\Theta}} \tau q_N \,\mathrm{d}a = 0.$$

the change of enthalpy H in time is modeled using the apparent heat approach

$$\dot{H} = c_{app}\dot{\vartheta}. \tag{8.83}$$

The phase change is thus assumed to be non-isothermal and the specific heat capacity differs in the temperature ranges according to (8.41)

$$c_{app} = \begin{cases} c_{s,f} & \text{if } \Theta < \Theta_s \\ 0.5(c_{s,f} + c_{f,g}) + L_{mel,vap}(\Theta_{liq,dew} - \Theta_{sol,bub})^{-1} & \text{if } \Theta_{sol,bub} \leq \Theta \leq \Theta_{liq,dew} \\ c_{f,g} & \text{if } \Theta > \Theta_{liq,dew} \end{cases} .$$

The discretized equation can also be formulated as a balance of powers that must be conserved at each node

$$P_I^h + P_I^q = \begin{cases} P_I^n & \text{boundary node} \\ 0 & \text{inner node} \end{cases} . \quad (8.84)$$

The contribution from the enthalpy follows directly from the apparent heat approach

$$P_I^h = \sum_{p=1}^{N_I^{inf}} N_{I,p} c_{app} \dot{\vartheta}_p m_p . \quad (8.85)$$

The term due to heat conduction is calculated using Fourier's law

$$P_I^q = -\sum_{p=1}^{N_I^{inf}} \mathbf{q}_p \cdot \frac{\partial N_{I,p}}{\partial \mathbf{x}} v_p, \quad \mathbf{q}_p = -k \sum_{I=1}^{N_p^{sup}} \vartheta_I \frac{\partial N_{I,p}}{\partial \mathbf{x}} . \quad (8.86)$$

In the OTM approach, the influence of recoil pressure and surface cooling is neglected. However, it can be implemented analogously to the previous section.

Two different concepts for modeling laser material interaction are investigated. The first approach assigns a power to each integration point below the laser employing a volumetric heat source model. The individual contribution is determined by the intensity function I consisting of a radial and axial factor $I(\mathbf{x}_p) = I_r(r) I_z(z)$

$$P_I^n = \sum_{p=1}^{N_I^{inf}} N_{I,p} \, P_{laser} \, I\left(\mathbf{x}_p\right) v_p . \quad (8.87)$$

In the second approach, the laser power is modeled using ray-tracing. A search algorithm determines the rays that interact with the surface. Each integration point possesses a detection sphere with a certain radius. If a ray lies in the sphere, the energy is assigned to that integration point (Fig. 8.16). Using only this concept the absorbed laser power at an integration point would depend on the discretization (Wessels et al. 2018a). To avoid this numerical artifact, a non-local formulation is employed. The laser power of a ray at the integration point is therefore weighted

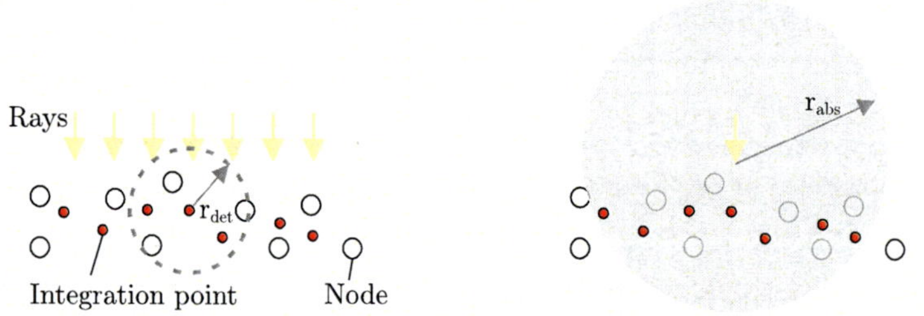

Fig. 8.16 (Left) Search algorithm to detect rays in contact with the material. (Right) Integration points within the influence of a ray defined by the absorption radius r_{abs}

with a normalized distance function. The scaling parameter depends on the number of integration points N_r^{abs} within the influence domain of a ray (Fig. 8.16). This approach assumes a penetration depth of the laser into the material. The Beer-Lambert model is employed to determine the weighting function

$$w_p^r = \frac{\exp\left(-\frac{\beta\|\mathbf{x}_r - \mathbf{x}_p\|}{r_{abs}}\right)}{Z}, \quad Z = \sum_{p=1}^{N_r^{abs}} \exp\left(-\frac{\beta\|\mathbf{x}_r - \mathbf{x}_p\|}{r_{abs}}\right). \tag{8.88}$$

The laser power at an integration point is computed by the weighted, absorbed energy of all rays N_p^{ray} belonging to the corresponding integration point

$$P_I^n = \sum_{p=1}^{N_I^{inf}} P_p^{ray}, \quad P_p^{ray} = \sum_{r=1}^{N_p^{ray}} \frac{(1-R_r)\,E_r}{\Delta t_r} w_p^r. \tag{8.89}$$

The calculation rule for the reflection R_r and the energy input E_r is given in Sect. 8.3.1.

8.6.2 *Temporal Discretization*

The discretized equations contain time derivatives. To avoid very small time steps an implicit integration scheme is preferred. The displacements at time step $n+1$ are computed using the Newmark method (Newmark 1959)

$$\begin{aligned} \mathbf{u}_{I\,n+1} &= \mathbf{u}_{I\,n} + \Delta t \mathbf{v}_{I\,n} + \frac{(\Delta t)^2}{2}\left[2\beta\dot{\mathbf{v}}_{I\,n+1} + (1-2\beta)\dot{\mathbf{v}}_{I\,n}\right] \\ \mathbf{v}_{I\,n+1} &= \mathbf{u}_{I\,n} + \Delta t\left[\gamma\dot{\mathbf{v}}_{I\,n+1} + (1-2\gamma)\dot{\mathbf{v}}_{I\,n}\right]. \end{aligned} \tag{8.90}$$

Stability can be ensured if the two numerical parameters are selected as $\beta = 0.25$ and $\gamma = 0.5$ leading to an amplitude error equal to zero. Thus, conservation of energy is guaranteed (Wood 1990). The derivative of temperature with respect to time is integrated using the implicit Euler method

$$\vartheta_{I\,n+1} = \vartheta_{I\,n} + \Delta t \dot{\vartheta}_{I\,n+1}. \tag{8.91}$$

Remark. In Chiumenti (1998) a lumped heat capacity matrix is recommended in (8.85) to avoid oscillations in the temperature field.

8.7 Computational Studies

Using a representative example, the effects of different modeling and discretization concepts can be displayed very clearly. Two spherical powder particles are irradiated by a laser. The geometry, the dimension, and the boundary conditions can be found in Fig. 8.17. The laser moves at a velocity of $v_{laser} = 0.2\,\text{m/s}$. The beam radius is 30 μm. Figure 8.18 shows the schematic sequence of the laser irradiation and the fusion behavior. The powder particles and the substrate are made of 316L stainless steel. The material parameters are taken from Company (1968) and are summarized in Table 8.1. The dependence of the heat conduction

$$k = \begin{cases} \left(9.345 + 0.0135\frac{1}{K}\Theta\right)\frac{W}{m\,K} & \Theta \leq \Theta_{mel} \\ 32.4\frac{W}{m\,K} & \Theta > \Theta_{mel} \end{cases}, \tag{8.92}$$

and the specific heat capacity on the temperature are assumed to be linear

$$c = \begin{cases} \left(376.406 + 0.258\frac{1}{K}\Theta\right)\frac{J}{K} & \Theta \leq \Theta_{mel} \\ 815\frac{J}{kg} & \Theta > \Theta_{mel} \end{cases}. \tag{8.93}$$

In this study, the apparent heat approach is used. Hence, the limit temperatures of a non-isothermal phase change need to be assumed. The transition range is chosen as 200 K. This leads to a solidus and liquidus temperature of $\Theta_{sol} = 1670\,\text{K}$ and $\Theta_{liq} = 1870\,\text{K}$. At the transition from liquid to gas phase the bubble and dew temperature is $\Theta_{bub} = 3030\,\text{K}$ and $\Theta_{dew} = 3230\,\text{K}$, respectively.

Two studies are carried out, comparing the influence of laser modeling and spatial discretization schemes. The third test investigates the plastic deformations that lead to residual stresses. In the end, an entire single track is simulated. With this tool, the optimal process parameters for Selective Laser Melting for any material can be determined virtually.

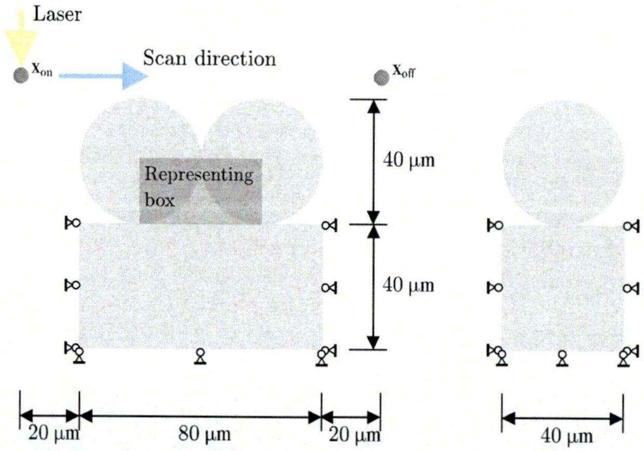

Fig. 8.17 Geometry, dimensions, and boundary conditions of the metal particle fusion test case. Laser irradiation starts at the position $\mathbf{x}_{on}$ and ends at $\mathbf{x}_{off}$

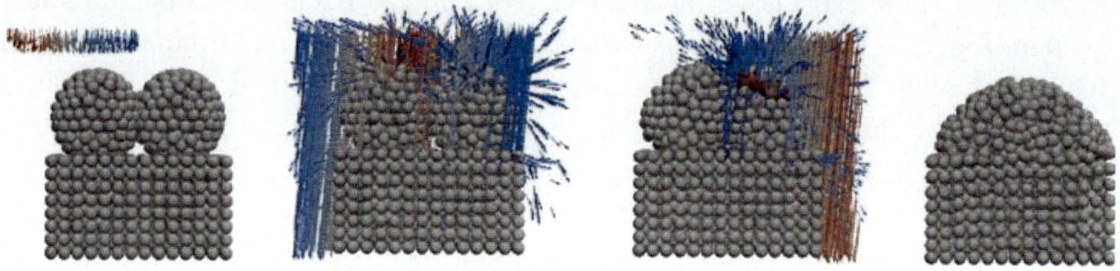

Fig. 8.18 Remelting of two spherical powder particles on a substrate using OTM method and ray-tracing

Table 8.1 Material properties of stainless steel 316L (r.t.: room temperature)

Thermomechanical parameter	Symbol	Value
Poisson's ratio (r.t)	ν	0.25
Compression modulus (r.t)	K	0.13 GPa
Viscosity of molten metal (r.t)	η	0.1 kg/(s·m)
Surface tension coefficient (r.t)	γ	1.78 N/m
Thermal expansion coefficient (r.t)	α_{th}	$1.75 \cdot 10^{-3}$
Latent heat of melting (vaporization)	L_{mel} (L_{vap})	0.299 MJ/kg (6.09 MJ/kg)
Melting (evaporating) temperature	Θ_{mel} (Θ_{vap})	1770 K (3130 K)
Initial density (r.t)	ρ_0	7900 kg/m^3

8.7.1 Comparison of Laser Modeling

The volumetric heat source assumes a certain penetration depth of the radiation into the body without any heat conduction. The model according to Gusarov et al. (2009) considers not only the penetration depth but also the heat transport via the reflected radiation inside the powder bed. The parameters for the optical thickness, the hemispherical reflectivity, and the maximum penetration depth are $\lambda = 2$, $\rho_h = 0.7$ and $L = 40\ \mu\text{m}$, respectively.

The second option is ray-tracing. The decomposition into rays enables the explicit determination of the propagation and reflection of the laser radiation. The time increment is selected to $\Delta t_r = 10^{-6}$ s. Each cross-section consists of $n_{rpl} = 844$ rays. The ratio of the magnetic permeability $\hat{\mu}$ and the refractive indices $\hat{n}$ between argon gas and stainless steel is $\hat{\mu} = 1$ and $\hat{n} = 2.4$.

Since the deformation of the substrate at the lower end is very small, only the upper part is modeled. To estimate the temperature gradient correctly, a heat flow condition is assumed at the lower edge. The heat transfer coefficient in (8.45) is selected to $h = 10^3$ W/(m^2 K).

The simulations are conducted using the OTM method to compare the two modeling concepts for laser-material interaction. The time increment of the implicit scheme is $\Delta t = 10^{-6}$ s. The parameter for stabilizing the configurational consistency (5.29) is set to $\epsilon = 10^{-8}$ and the parameter for the LME shape functions to $\gamma = 1.2$. The influence of Marangoni convection, recoil pressure, and plastic deformations are neglected in this study.

A look at the graphical evaluation in Fig. 8.19 shows the different temperature distributions in the body. Using a volumetric heat source, the temperature gradually decreases over the height of the particles. With ray-tracing, the concentration of heat and thus the temperature at the surface is significantly higher. A study of the maximum temperature in the body during the melting process (Fig. 8.20) shows that ray-tracing leads to higher values. Therefore, the two powder particles fuses together

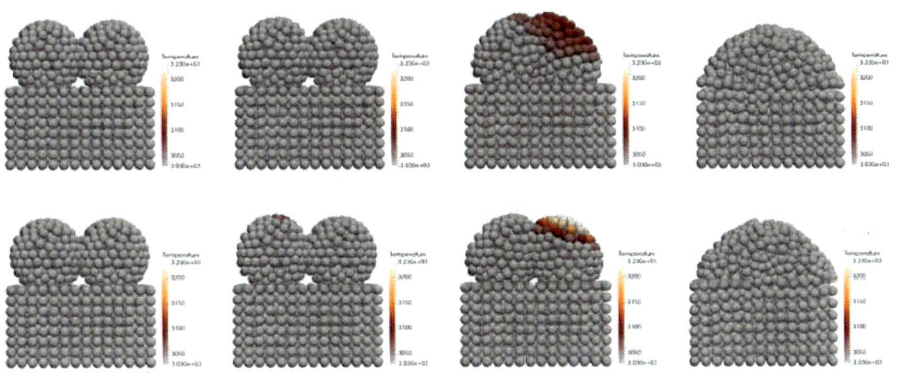

Fig. 8.19 Remelting of two metal particles with a substrate. (Upper line) Volumetric heat source. (Lower line) Ray-tracing

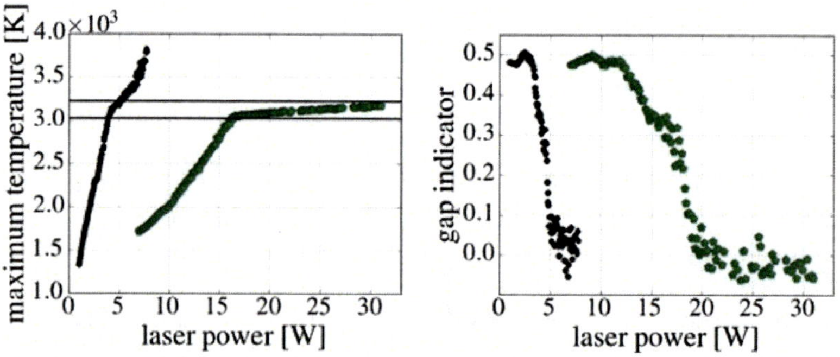

Fig. 8.20 Comparison ray-tracing (black dots) and volumetric heat source (green pentagons) in dependence on the laser power. (Left) Maximum temperature in the melt. (Right) Amount of Argon gas in the representative box indicating the remelt behavior

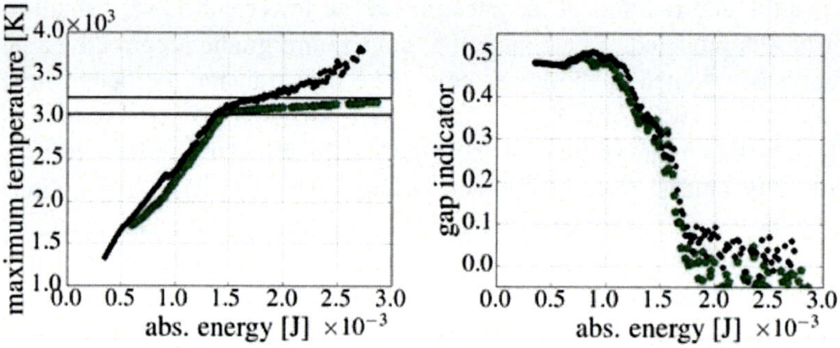

Fig. 8.21 Comparison ray-tracing (black dots) and volumetric heat source (green pentagons) in dependence on the absorbed energy. (Left) Maximum temperature in the melt. (Right) Amount of argon gas in the representative box indicating the remelt behavior

at a lower laser power (Fig. 8.20). To measure fusion, the amount of argon gas in a representative box (Fig. 8.17) is calculated. This value is based on the ratio of the material v_{metal} to the complete volume of the box v_{box}

$$\Pi = 1 - \frac{v_{metal}}{v_{box}}. \tag{8.94}$$

Initially, the percentage of argon gas is 48%.

If the absorbed heat is considered instead of the laser power, the curves between the two modeling approaches almost coincide (Fig. 8.21). Thus, both modeling concepts reproduce the same physical behavior. Only the maximum temperature is slightly higher in the case of ray-tracing because the heat is more concentrated at the surface. The ratio of laser power to absorbed energy is different in both methods (Fig. 8.22) leading to the different outcomes in Fig. 8.20.

Conclusion. Volumetric heat sources are designed for laser irradiation on straight surfaces. This is not the case when simulating the fusion of powder particles. Thus,

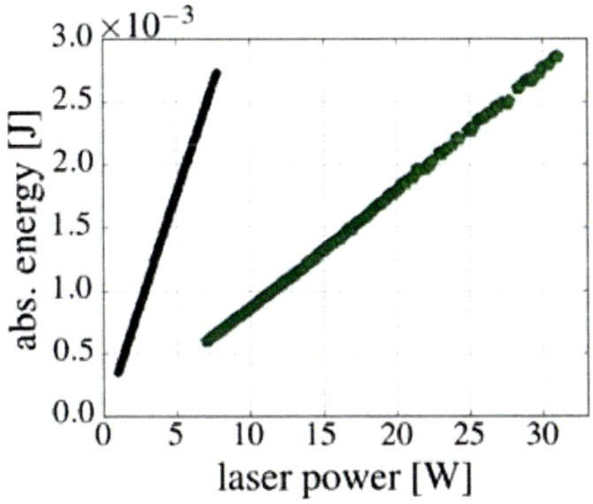

Fig. 8.22 Absorbed energy as a function of laser power using ray tracing (black dots) and a volumetric heat source (green pentagons)

a part of the heat is not assigned to the material but to the surrounding argon gas. This explains the difference between absorbed heat and laser power in both modeling concepts. The shape of the surface must be included in the formulation of a volumetric heat source. Although the ray-tracing scheme increases the computational effort, the change in the surface is automatically taken into account. In addition, fewer material parameters are required compared to a volumetric heat source, since the reflection is considered intrinsically.

Remark. The exit of heat at the lower end determines the amount of heat remaining in the system. This value is a crucial parameter in SLM and has to be modeled accurately. As shown in Wessels et al. (2018b), the melting behavior depends strongly on the heat flow through the substrate.

8.7.2 Comparison of Solution Schemes

The second study shows a comparison between two spatial discretization schemes. The example from the previous section is used again. The dimensions of the substrate layer are $100 \times 60 \times 120\,\mu m$ (length × width × height). The ambient temperature is set to 823.1 K. Both approaches use the volumetric heat source from the previous example. The laser power is 9.2 W and the scan rate 0.2 m/s.

The differences between the two approaches are only the time integration scheme and the modeling of the transition from solid to liquid phase. With SPH, the particles in the solid phase are assumed to be rigid. Since the laser melts the material very quickly, there are no large deformations in the solid phase and this assumption has no major influence on the deformation behavior during fusion. The comparison in Fig. 8.23 shows strong differences between the two discretization schemes. The material liquefies much more when the equations are discretized using SPH. When

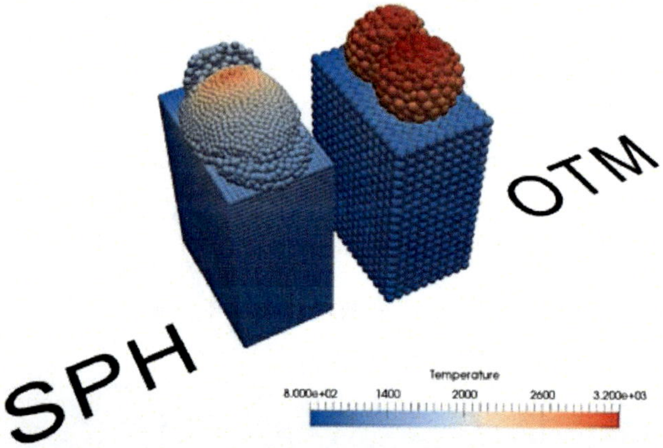

Fig. 8.23 Temperature distribution during melting. Comparison between SPH and the OTM method

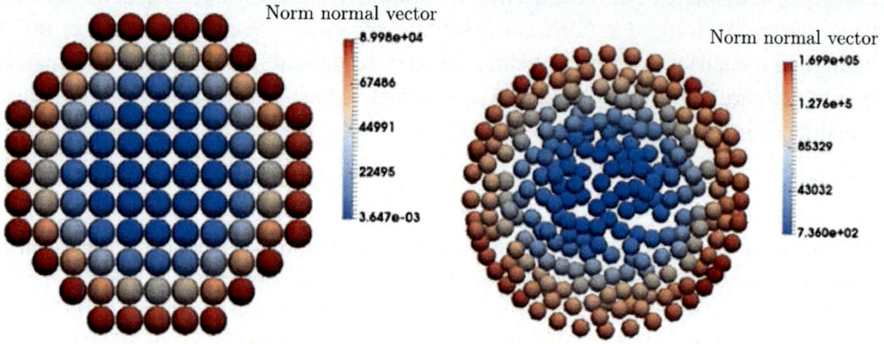

Fig. 8.24 Norm of the resulting normal vectors. (Left) Discretization based on SPH. (Right) Discretization based on the OTM method

the OTM method is applied, the geometry of the two spheres is visibly preserved for longer.

Conclusion. The big weakness of the OTM method without any additional correction schemes is the non-fulfillment of the integration constraint. The resulting normal vectors from this requirement are used to identify the surface nodes. The curvature is also calculated from these vectors. Thus, a fulfillment of the integration constraint is essential. As can be seen in Fig. 8.24 the SPH predicts the resulting normal vectors better. In addition, the LME shape functions used in the OTM method show a too stiff response behavior. Although the factor to stabilize the negative effects due to a violation of the configurational consistency is very small, this additional term also leads to an unphysical behavior.

8.7.3 Residual Stresses

The presented approach in the context of the OTM method directly models the phase transition from solid to liquid. This allows the computation of residual stresses caused by melting and cooling. The example from the previous sections is considered again. The parameters for the plasticity model in (8.37) and (8.38) are $Y = 206.85$ MPa, $H = 413.7$ MPa and $\mu_p = 1$ MPa/s. In addition, the dependence of the Young's modulus

$$E = \begin{cases} (225.916 - 0.092\,\Theta/K) & GPa\ 300\,K \leq \Theta \leq \Theta_{sol} \\ (706.884 - 0.392605\,\Theta/K) & GPa\ \Theta_{sol} \leq \Theta \leq \Theta_{liq} \\ 0.195 & GPa\ \Theta > \Theta_{liq} \end{cases} \tag{8.95}$$

and the coefficient of thermal expansion on the temperature is considered. For the latter, the reference temperature Θ_0 must be adjusted accordingly

$$(\alpha, \Theta_0) = \begin{cases} (1.4604 \cdot 10^{-5}\,1/K, 300\,K) & 300\,K \leq \Theta \leq \Theta_{mel} \\ (4.4235 \cdot 10^{-5}\,1/K, 1700\,K) & \Theta_{mel} \leq 2000\,K \\ (1.9218 \cdot 10^{-5}\,1/K, 2000\,K) & \Theta > 2000\,K \end{cases} . \tag{8.96}$$

The full height of the substrate of 120 μm is considered. The ambient temperature is 823.15 K. The approach due to Gusarov et al. (2009) is used to model the laser entry into the material. All remaining parameters are equal to the previous examples. As can be seen in Fig. 8.25, plastic deformations occur at the boundary between the melted and unmelted areas. This effect was also observed experimentally in Mercelis and Kruth (2006).

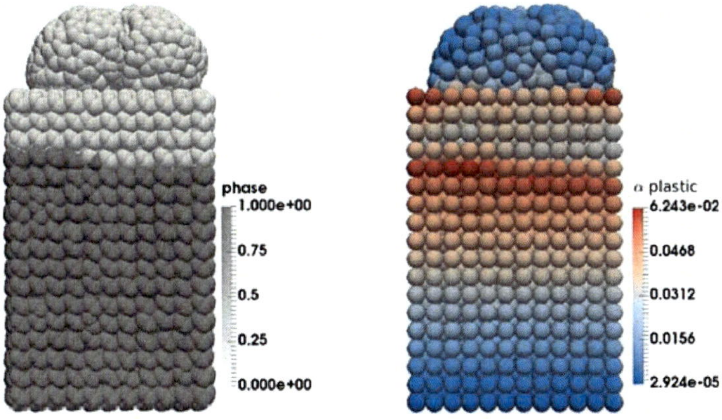

Fig. 8.25 Remelting of two powder particles with a substrate. (Left) Indication of melted and unmelted nodes. (Right) Nodal accumulated plastic strains

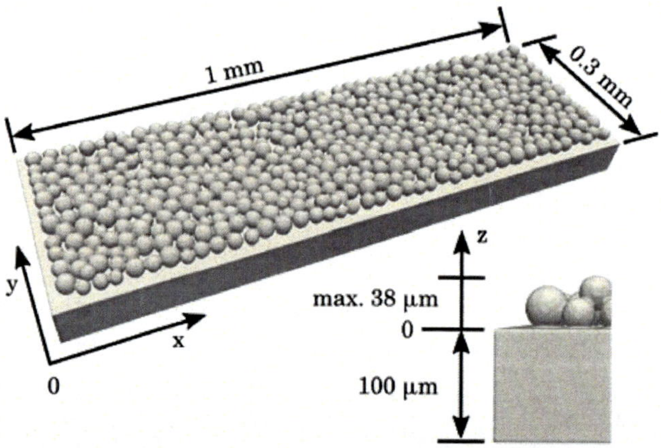

Fig. 8.26 Geometry and dimensions of the fully resolved single track

8.7.4 Single Track Computations

The SPH seems to lead to physically more meaningful results compared to the OTM method. To underline this assumption, the results of a single track are compared with experimental observations. With the help of a DEM code, the particles are deposited on a metal plate. The geometry and dimensions are shown in Fig. 8.26. Except for the upper surface, the displacement perpendicular to the metal plate is fixed at all sides. The mean particle diameter is 27 µm and the standard deviation is $\sigma = 4.25$ µm. At the bottom of the substrate, the temperature is assumed to be 823.15 K. The distribution of the laser radiation in the cross-section corresponds to (8.18). The radius of the laser beam is 35 µm. The absorption of the laser radiation is modeled using the Beer-Lambert approach (8.25). The penetration depth is $L = 28$ µm and the extinction coefficient $\beta = 0.3$. The laser moves centrally along the x-direction from x = 0.1 mm to x = 0.9 mm. When modeling the single track, the specific enthalpy is selected as the solution function. Due to phase change, the relationship to the temperature divides into three ranges

$$H = \begin{cases} (136141.8 + 892.406\,\Theta/K) & J/K \;\; 300\,K \leq \Theta \leq \Theta_{mel} \\ (1684510,2 + 815\,\Theta/K) & J/K \;\; \Theta_{mel} \leq \Theta_{vap} \\ 815\,\Theta/K & J/K \;\; \Theta > \Theta_{vap} \end{cases} . \tag{8.97}$$

To model the Marangoni convection, the dependence of the surface tension on the temperature must be taken into account

$$\gamma = \begin{cases} 1.78 & N/m \;\; 300\,K \leq \Theta \leq \Theta_{mel} \\ 1.78 - 0.00089\,\Theta/K & N/m \;\; \Theta > \Theta_{mel} \end{cases} . \tag{8.98}$$

Comparison with experiments. In Yadroitsev et al. (2010), a layer of stainless steel 316L is irradiated with a laser power of 50 W. Different scan rates lead to different

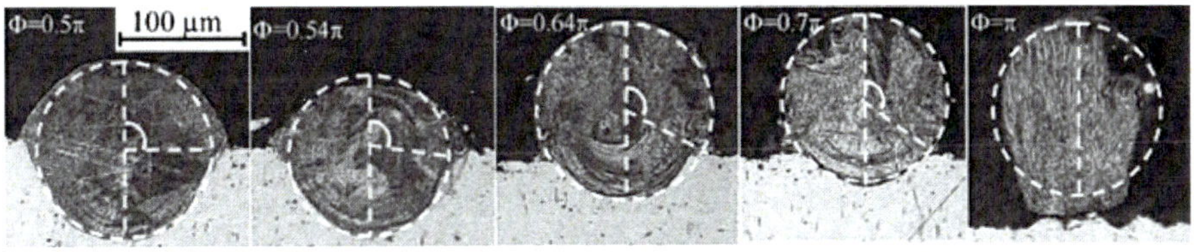

Fig. 8.27 Slices of remelted vectors (Yadroitsev et al. 2010). Laser power 50 W. Scan rates 0.06, 0.08, 0.12, 0.16, 0.2 m/s (left to right)

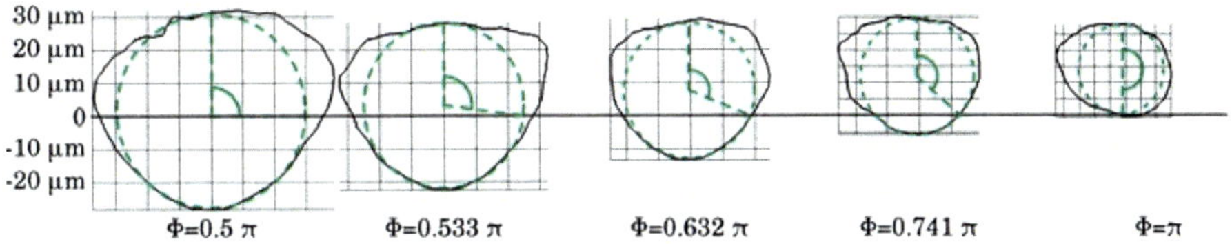

Fig. 8.28 Comparison simulation (black line) with experiments (dashed green line) from Fig. 8.27

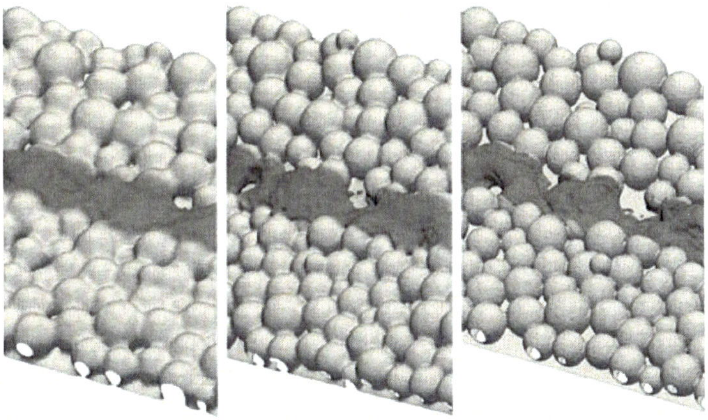

Fig. 8.29 Snapshots of the remelted vector in the same interval showing the increasing detail of the balling effect with discretizations of 3 μm (left), 2 μm (middle), 1 μm (right)

penetration depths and a different radii of the vector (Fig. 8.27). The comparison with the simulation (Fig. 8.28) shows that the occurring phenomena can be reproduced qualitatively well using the presented SPH algorithm. Analogous to the experiment, the penetration depth is reduced by an increase of the scan rate. The remelted material also possesses a cylindrical shape.

A proper resolution of the domain is decisive for accurate numerical results. A sufficient number of particles must be chosen for the solution to be in the convergent range. Investigations in Fürstenau et al. (2019b) show that a distance of 3 μm between the particles in the initial state provides an almost convergent behavior. Negative effects, like balling, can be recognized, which are not visible with a too coarse resolution (Fig. 8.29).

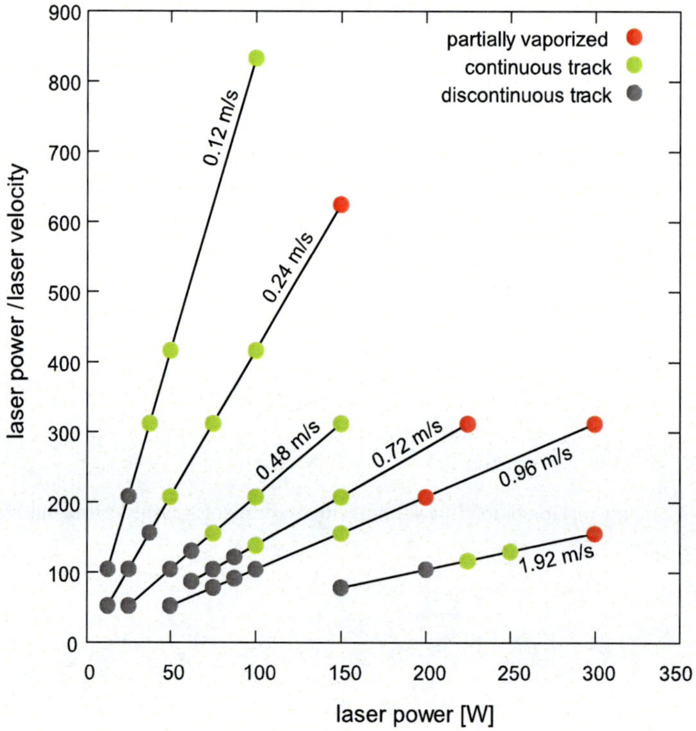

Fig. 8.30 Process map generated with simulated data using SPH for stainless steel 316L

Virtual process maps. The design of the printing process requires knowledge of the optimum process parameters. For this purpose, process maps are usually created that indicate the laser power and scan rate range in which 3D printing can be carried out for a specific material on the corresponding machine. To achieve the desired result, the remelted vector must have a cylindrical shape in a cross-section that has penetrated the underlying layer. The process parameters in the green area on the process map guarantee optimal 3D printing. The creation of these maps purely experimentally consumes a lot of material, time, and energy. One advantage of explicit time integration schemes is the efficient computation of large dimensional applications on GPUs. Using the in-house code ADDITIVE, a complete single trace with a resolution of 3 μm (1 million particles) is computed in about half an hour on an Nvidia GTX Titan Black GPU. This allows the generation of virtual process maps (Fig. 8.30) with acceptable computation time. The permissible range of laser power and scan rate can now be determined purely virtually for each material (Fig. 8.30). This saves time, material, and energy.

Influence of recoil pressure. In the previous studies the influence of recoil pressure was assumed to be negligible. An additional investigation shows the influence of

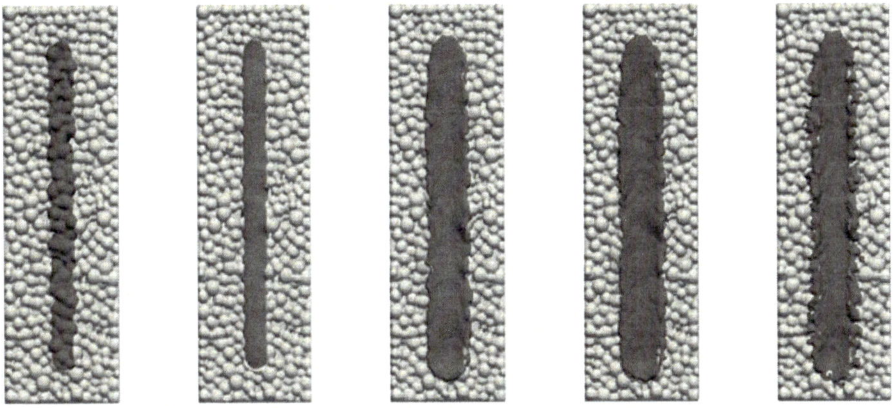

Fig. 8.31 Remelted vector (dark gray). Added physical effects from left to right: Only gravity, plus surface tension, plus Marangoni convection, plus recoil pressure, plus surface cooling

individual physical effects on the behavior of the melt pool. If Marangoni convection and surface tension are neglected in addition to recoil pressure and associated cooling effect, no melting of the particles occurs. Only if the surface tension is included the powder particles fuse together (Fig. 8.31). That demonstrates that this effect is the main driver in Selective Laser Melting. If Marangoni convection is taken into account, the melt pool widens due to the outward surface force. In addition, the remelted area is indented. The recoil pressure intensifies this effect. If the contribution from surface cooling is additionally taken into account, the width of the remelted area is reduced. At the same time its height increases. The parameters in the calculation of the recoil pressure (8.50) and the surface cooling are the enthalpy of vaporization $\triangle H_{vap} =$ 6, 09 MJ/kg, the molar mass $M = 55.85$ g/mol, the gas constant $R = 8.314\,\mathrm{kg\,m^2/(s^2\,mol\,K)}$ and the sticking coefficient, which can be assumed to be one for metals, i.e. $A \approx 1$.

Conclusion. The SPH formulation, presented in Sect. 8.5, meets the requirements on conservation of balance equations. The integration constraint is only slightly violated. The radius of the search algorithm is the same for each particle. Since no tensile stresses occur, the tensile instability plays no role. In SPH, the conservation of the equilibrium of forces during the redefinition of the neighborhoods is mostly not tested as well as the influence from under-integration. The shape functions do not fulfill the reproducing conditions. Nevertheless, the results show a response behavior that agrees qualitatively well with experimental outcomes. The OTM method fulfills the reproducing conditions and conserves momentum and angular momentum. However, the integration constraint is strongly violated. Due to its importance for the calculation, especially for the forces at the surface, the SPH is preferable compared to the OTM method at the current stage of development.

Remark 1 The comparison of the simulation results with experimental data shows differences in the diameters of the printed vectors. In the experiment, the laser irradiates a powder bed with a height of 50 μm. The average particle diameter is 11.2 μm and thus smaller than the diameter in the simulation. For computational reasons, it was at that time not possible to carry out simulations with smaller particle diameters. It is also important to check whether the influences from recoil pressure may be neglected.

Remark 2 In meshfree methods, only the position vectors are given to represent the geometry. For a better representation and evaluation, algorithms are available which determine a surface from a set of points. Especially with the visualization of powder geometries, it must be paid attention that the surfaces are reproduced accurately by the algorithm.

References

A.W. Adamson, A.P. Gast, *Physical Chemistry of Surfaces*, 6th edn. (Wiley Interscience, New York, 1997)

S.I. Anisimov, V.A. Khokhlov *Instabilities in Laser-Matter Interaction* (CRC press, 1995)

A. Bauereiß, T. Scharowsky, C. Körner, Defect generation and propagation mechanism during additive manufacturing by selective beam melting. J. Mater. Process. Technol. **214**(11), 2522–2528 (2014)

D. Bäuerle, *Laser Processing and Chemistry* (Springer Science & Business Media, 2013)

T. Belytschko, R.L. Chiapetta, H.D. Bartel, Efficient large scale non-linear transient analysis by finite elements. Int. J. Numer. Methods Eng. **10**(3), 579–596 (1976)

P. Bidare, I. Bitharas, R.M. Ward, M.M. Attallah, A.J. Moore, Fluid and particle dynamics in laser powder bed fusion. Acta Materialia **142**, 107–120 (2018)

C. Bonacina, G. Comini, A. Fasano, M. Primicerio, Numerical solution of phase-change problems. Int. J. Heat Mass Transf. **16**(10), 1825–1832 (1973)

E.Y. Bormashenko, *Physics of Wetting: Phenomena and Applications of Fluids on Surfaces* (Walter de Gruyter GmbH & Co KG, 2017)

S. Chandrasekhar, *Hydrodynamic and Hydromagnetic Stability* (Courier Corporation, 2013)

T.H.C. Childs, C. Hauser, M. Badrossamay, Selective laser sintering (melting) of stainless and tool steel powders: experiments and modelling. Proc. Inst. Mech. Eng. Part B: J. Eng. Manuf. **219**(4), 339–357 (2005)

M. Chiumenti, *Constitutive Modeling and Numerical Analysis of Thermo-Mechanical Phase Change Systems*. Ph.D. thesis, UPC BarcelonaTech, Spain (1998)

P.W. Cleary, J.J. Monaghan, Conduction modelling using Smoothed Particle Hydrodynamics. J. Comput. Phys. **148**(1), 227–264 (1999)

The International Nickel Company, editor, *Mechanical and Physical Properties of the Austenitic Chromium-Nickel Stainless Steels at elevated Temperatures*, 3rd edn. (INCO Databooks, New York, 1968)

R. Courant, K. Friedrichs, H. Lewy, Über die partiellen Differenzengleichungen der mathematischen Physik. Mathematische Annalen **100**(1), 32–74 (1928)

A.J.C. Crespo, J.M. Domínguez, B.D. Rogers, M. Gómez-Gesteira, S. Longshaw, R. Canelas, R. Vacondio, A. Barreiro, O. García-Feal, DualSPHysics: open-source parallel CFD solver based on Smoothed Particle Hydrodynamics (SPH). Comput. Phys. Commun. **187**, 204–216 (2015)

S. Dietrich, M. Wunderer, A. Huissel, M.F. Zaeh, A new approach for a flexible powder production for additive manufacturing. Procedia Manuf. **6**, 88–95 (2016)

EPMA. Introduction to Additive Manufacturing technology: a guide for designers and engineers (2016)

P. Fischer, V. Romano, H.P. Weber, N.P. Karapatis, E. Boillat, R. Glardon, Sintering of commercially pure titanium powder with a Nd: YAG laser source. Acta Materialia **51**(6), 1651–1662 (2003)

J.P. Fürstenau, C. Weißenfels, P. Wriggers, Free surface tension using ISPH. Comput. Mech. **95**, 175–187 (2019)

J.P. Fürstenau, H. Wessels, C. Weißenfels, P. Wriggers, Generating virtual process maps of SLM using powder scale SPH simulations. Comput. Part. Mech. **339**, 91–114 (2019)

R. Ganeriwala, T.I. Zohdi, A coupled discrete element-finite difference model of Selective Laser Sintering. Granul. Matter **18**(2) (2016). ISSN 1434-5021

J. Goldak, A. Chakravarti, M. Bibby, A new finite element model for welding heat sources. Metall. Trans. B **15**(2), 299–305 (1984)

J.A. Goldak, M. Akhlaghi, *Computational Welding Mechanics* (Springer, New York, 2005)

D.D. Gu, W. Meiners, K. Wissenbach, R. Poprawe, Laser additive manufacturing of metallic components: materials, processes and mechanisms. Int. Mater. Rev. **57**(3), 133–164 (2012)

V. Gunenthiram, P. Peyre, M. Schneider, M. Dal, F. Coste, R. Fabbro, Analysis of laser-melt pool-powder bed interaction during the selective laser melting of a stainless steel. J. Laser Appl. **29**, 022303 (2017)

A.V. Gusarov, I. Yadroitsev, P. Bertrand, I. Smurov, Model of radiation and heat transfer in laser-powder interaction zone at Selective Laser Melting. J. Heat Transf. **131**(7), 072101 (2009)

A.V. Gusarov, M. Pavlov, I. Smurov, Residual stresses at laser surface remelting and additive manufacturing. Phys. Procedia **12**, 248–254 (2011)

R.J. Hebert, Metallurgical aspects of powder bed metal additive manufacturing. J. Mater. Sci. **51**(3), 1165–1175 (2016)

J. Hecht, *Understanding Lasers: An entry level Guide* (Wiley-IEEE Press, 2019)

M. Hirschler, G. Oger, U. Nieken, D. Le Touzé, Modeling of droplet collisions using smoothed particle hydrodynamics. Int. J. Multiph. Flow **95**, 175–187 (2017)

T.R.J. Hughes, *The Finite Element Method* (Prentice Hall, Englewood Cliffs, New Jersey, 1987)

Q. Jiang, Z. Wen, *Thermodynamics of Materials* (Springer Science & Business Media, 2011)

S.A. Khairallah, A.T. Anderson, A. Rubenchik, W.E. King, Laser powder-bed fusion additive manufacturing: physics of complex melt flow and formation mechanisms of pores, spatter, and denudation zones. Acta Materialia **108**, 36–45 (2016)

W.E. King, H.D. Barth, V.M. Castillo, G.F. Gallegos, J.W. Gibbs, D.E. Hahn, C. Kamath, A.M. Rubenchik, Observation of keyhole-mode laser melting in laser powder-bed fusion additive manufacturing. J. Mater. Process. Technol. **214**(12), 2915–2925 (2014)

W.E. King, A.T. Anderson, R.M. Ferencz, N.E. Hodge, C. Kamath, S.A. Khairallah, A.M. Rubenchiks, Laser powder bed fusion Additive Manufacturing of metals; physics, computational, and materials challenges. Appl. Phys. Rev. **2**(4), 041304 (2015)

A. Klassen, T. Scharowsky, C. Körner, Evaporation model for beam based additive manufacturing using free surface lattice Boltzmann methods. J. Phys. D: Appl. Phys. **47**(27), 275303 (2014)

C. Kleinstreuer, *Engineering Fluid Dynamics: An Interdisciplinary Systems Approach* (Cambridge University Press, 1997)

J. Korelc, S. Stupkiewicz, Closed-form matrix exponential and its application in finite-strain plasticity. Int. J. Numer. Methods Eng. **98**(13), 960–987 (2014)

J.P. Kruth, L. Froyen, J. van Vaerenbergh, P. Mercelis, M. Rombouts, B. Lauwers, Selective laser melting of iron-based powder. J. Mater. Process. Technol. **149**(1–3), 616–622 (2004)

A. Lawley, Preparation of metal powders. Ann. Rev. Mater. Sci. **8**(1), 49–71 (1978)

C. Limmaneevichitr, S. Kou, Experiments to simulate effect of Marangoni convection on weld pool shape. Weld. J. **79**(8), 231–237 (2000)

S.C.H. Lu, K.S. Pister, Decomposition of deformation and representation of the free energy function for isotropic thermoelastic solids. Int. J. Solids Struct. **11**(7–8), 927–934 (1975)

J. Lubliner, *Plasticity Theory* (Macillan, London, 1990)
M.J. Matthews, G. Guss, S.A. Khairallah, A.M. Rubenchik, P.J. Depond, W.E. King, Denudation of metal powder layers in laser powder bed fusion processes. Acta Materialia **114**, 33–42 (2016)
P. Mercelis, J.P. Kruth, Residual stresses in selective laser sintering and selective laser melting. Rapid Prototyp. J. **12**(5), 254–265 (2006)
J.J. Monaghan, A. Kos, Solitary waves on a cretan beach. J. Waterw. Port Coast. Ocean Eng. **125**(3), 145–155 (1999)
N.M. Newmark, A method of computation for structural dynamics, in *Proceedings of ASCE, Journal of Engineering Mechanics* (American Society of Civil Engineers, 1959), pp. 557–587
W. Polifke, *Wärmeübertragung: Grundlagen, analytische und numerische Methoden* (Pearson Deutschland GmbH, 2009)
J.K. Prescott, R.A. Barnum, On powder flowability. Pharm. Technol. **24**(10), 60–85 (2000)
A.M. Prokhorov, *Laser Heating of Metals* (CRC Press, 2018)
C. Qiu, C. Panwisawas, M. Ward, H.C. Basoalto, J.W. Brooks, M.M. Attallah, On the role of melt flow into the surface structure and porosity development during selective laser melting. Acta Materialia **96**, 72–79 (2015)
R. Rai, J.W. Elmer, T.A. Palmer, T. DebRoy, Heat transfer and fluid flow during keyhole mode laser welding of tantalum, Ti-6Al-4V, 304L stainless steel and vanadium. J. Phys. D Appl. Phys. **40**(18), 5753 (2007)
P.W. Randles, L.D. Libersky, Smoothed Particle Hydrodynamics: some recent improvements and applications. Comput. Methods Appl. Mech. Eng. **139**(1–4), 375–408 (1996)
V. Semak, A. Matsunawa, The role of recoil pressure in energy balance during laser materials processing. J. Phys. D Appl. Phys. **30**(18), 2541 (1997)
M. Shiomi, K. Osakada, K. Nakamura, T. Yamashita, F. Abe, Residual stress within metallic model made by selective laser melting process. CIRP Ann. **53**(1), 195–198 (2004)
J.C. Simo, *Numerical Analysis and Simulation of Plasticity*. Handbook of Numerical Analysis VI. Elsevier Science B.V. (1998)
W.M. Steen, J. Mazumder, *Laser Material Processing*, 6th edn. (Springer science & Business media, 2010)
L. Thijs, F. Verhaeghe, T. Craeghs, J. Van Humbeeck, J.P. Kruth, A study of the microstructural evolution during selective laser melting of Ti-6Al-4V. Acta Materialia **58**(9), 3303–3312 (2010)
N.K. Tolochko, S.E. Mozzharov, I.A. Yadroitsev, T. Laoui, L. Froyen, V.I. Titov, M.B. Ignatiev, Balling processes during selective laser treatment of powders. Rapid Prototyp. J. **10**(2), 78–87 (2004)
M. von Allmen, A. Blatter, *Laser-beam Interactions with Materials: Physical Principles and Applications*, vol. 2 (Springer Science & Business Media, 2013)
D. Wang, S. Wu, F. Fu, S. Mai, Y. Yang, Y. Liu, C. Song, Mechanisms and characteristics of spatter generation in SLM processing and its effect on the properties. Mater. Des. **117**, 121–130 (2017)
C. Weißenfels, Direct nodal imposition of surface loads using the divergence theorem. Finite Elem. Anal. Des. **165**, 31–40 (2019)
H. Wessels, *Thermo-Mechanical Modeling for Selective Laser Melting*. Ph.D. thesis, Leibniz Universität Hannover, Germany (2019)
H. Wessels, T. Bode, C. Weißenfels, P. Wriggers, T.I. Zohdi, Investigation of heat source modeling for selective laser melting. *Computational Mechanics*, pp. 1–22 (2018a)
H. Wessels, C. Weißenfels, P. Wriggers, Metal particle fusion analysis for additive manufacturing using the stabilized optimal transportation meshfree method. Comput. Methods Appl. Mech. Eng. **339**, 91–114 (2018b)
P.J. Withers, H.K.D.H. Bhadeshia, Residual stress. Part 2. Nature and origins. Mater. Sci. Technol. **17**(4), 366–375 (2001)
W.L. Wood, *Practical Time-Stepping Schemes*, vol. 6 (Clarendon Press, Oxford, 1990)
I. Yadroitsev, *Selective Laser Melting: Direct manufacturing of 3D-objects by Selective Laser Melting of Metal Powders* (Lambert Academic Publishing, 2009)

I. Yadroitsev, I. Smurov, Surface morphology in selective laser melting of metal powders. Phys. Procedia **12**, 264–270 (2011)

I. Yadroitsev, A. Gusarov, I. Yadroitsava, I. Smurov, Single track formation in selective laser melting of metal powders. J. Mater. Process. Technol. **210**(12), 1624–1631 (2010)

C.Y. Yap, C.K. Chua, Z.L. Dong, Z.H. Liu, D.Q. Zhang, L.E. Loh, S.L. Sing, Review of selective laser melting: materials and applications. Appl. Phys. Rev. **2**(4), 041101 (2015)

T. Young, III. An essay on the cohesion of fluids. Philos. Trans. R. Soc. Lond. **95**, 65–87 (1805)

T.I. Zohdi, Additive particle deposition and selective laser processing-a computational manufacturing framework. Comput. Mech. **54**(1), 171–191 (2014)

T.I. Zohdi, *Dynamics of Charged Particulate Systems: Modeling, Theory and Computation* (Springer Science & Business Media, 2012)

T.I. Zohdi, Rapid simulation of laser processing of discrete particulate materials. Arch. Comput. Methods Eng. **20**(4), 309–325 (2013)

Chapter 9
Modeling Direct Poly Printing

Photopolymerization is based on a chemical reaction. The material is initially in liquid to viscous form. As described in Chap. 2, different technologies for printing polymers exist. For the modeling, the process plays a minor role, since only the material parameters and the geometric structure differ. This chapter concentrates on the Additive Manufacturing process according to Stieghorst (2017). However, the modeling approach can also be applied to stereolithography or polyjet printing.

In Direct Poly Printing (DPP), the curing process is locally controlled by a laser to print the desired line width and height. The concept of Stieghorst (2017) was primarily developed to fabricate patient-specific, flexible neural implants based on biocompatible silicone. With DPP, the viscous material is deposited on a substrate or already printed layers via a mobile dispenser. An accompanying laser irradiates the material directly after placement (Fig. 9.1).

Analogous to Chap. 8, first the main physical phenomena and the most important experimental results are described. Afterward, the modeling of the laser-material interaction is introduced. The differential equations are discretized using Peridynamics. The numerical evaluations investigate the influence of the laser power and the extrusion velocity on the printing process.

9.1 Physical Phenomena

Many physical effects that occur during Selective Laser Melting also influence the printing of polymers. A laser irradiates the polymer and the absorbed heat is transported within the material and simultaneously leads to an increase in temperature in the body. In contrast to SLM, there is no phase change. The absorbed heat influences a chemical reaction that hardens the material. The laser power must be kept

C. Weißenfels, *Simulation of Additive Manufacturing using Meshfree Methods*, Lecture Notes in Applied and Computational Mechanics 97,
https://doi.org/10.1007/978-3-030-87337-0_9

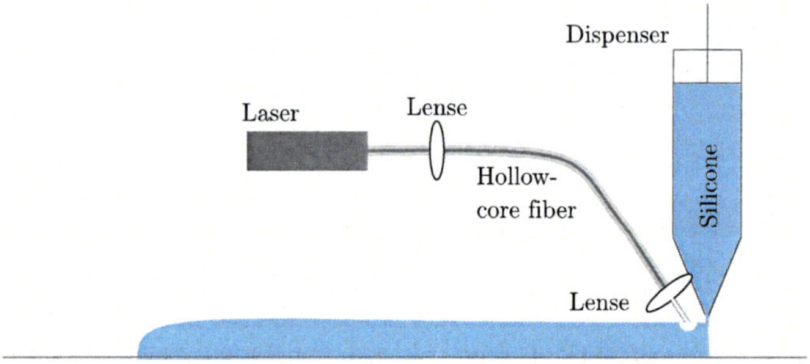

Fig. 9.1 Schematic description of Direct Poly Printing (DPP)

low so that no burns occur at the surface of the polymer. Hence, the vaporization of material can be excluded. The process aims to harden the material right after the extrusion. Hence, movements in the liquid material due to Marangoni convection are negligible. Especially the control of the viscosity is crucial for a successful print. In this section, the most important physical effects are listed, whereby analogous phenomena to Sect. 8.1 are no longer described in detail.

9.1.1 *Laser Matter Interaction*

In contrast to metals, polymers are semi-transparent. In general, this can lead to a transmission of the laser radiation through the material. The energy of the laser divides into absorption a, reflection r and transmission t

$$a(\lambda, \Theta) + r(\lambda, \Theta) + t(\lambda, \Theta) = 1. \tag{9.1}$$

The individual coefficients depend not only on the material but also on the wavelength of the laser radiation λ and the temperature Θ.

9.1.2 *Heat Transfer*

The depositing process does not result in a large mixing of the material in the liquid phase. Thus, convection plays almost no role and conduction dominates the heat transport. The emission of heat via the surface by radiation can also be neglected.

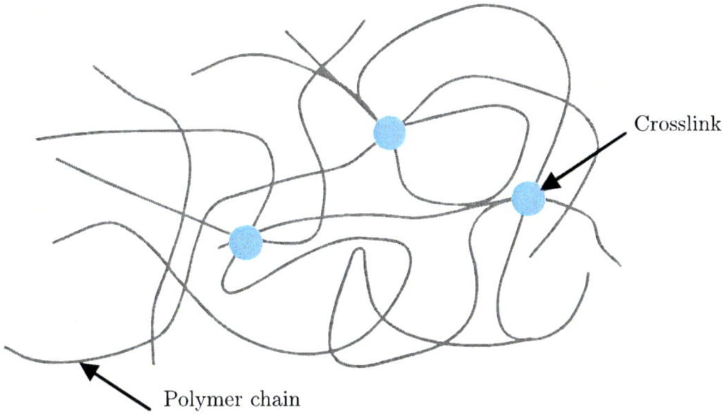

Fig. 9.2 Molecular structure: Cross-linked polymer chains

9.1.3 Heat Capacity and Crosslinking

Analogous to Selective Laser Melting, the absorbed energy leads to a temperature change in the body. Unlike metals, in photopolymerization, energy is not converted into latent heat to change phase. Instead, a chemical reaction takes place, which ensures the transition from liquid to solid. In the liquid form, the molecular chains are mobile. The chemical reaction cross-links the molecular chains (Fig. 9.2) and the stiffness increases. Elastomers have a wide meshed structure whereas thermosets a close-meshed. Depending on the polymer, different types of cross-linking occur (Elias 1996). When biocompatible silicone is printed, platinum-catalyzed cross-linking of the silicone polymers takes place (Stieghorst 2017). In this case, cross-linking depends on the temperature, the ratio of cross-linker to starting polymer, the concentration of platinum catalysts, and inhibitors, which can accelerate or slow down reactions. In addition, heat is released during cross-linking.

9.1.4 Surface Tension and Wetting

The dimensions of the individual layers must be small to print complex geometries. Since the material is initially in a liquid state, capillary action must be taken into account. During curing the influence of this process reduces. When it comes to over-printing, a certain spreading must be ensured so that the first layer can be completely wetted. The value of γ at which complete wetting is no longer possible is called critical surface tension. With silicone rubber, the disperse portion γ^d determines the surface tension. The polar part γ^p is negligibly small. The surface tension of silicone rubber is about $20 \cdot 10^{-3}$ N/m and lower than the critical value of $24 \cdot 10^{-3}$ N/m. Thus, a complete wetting can be realized.

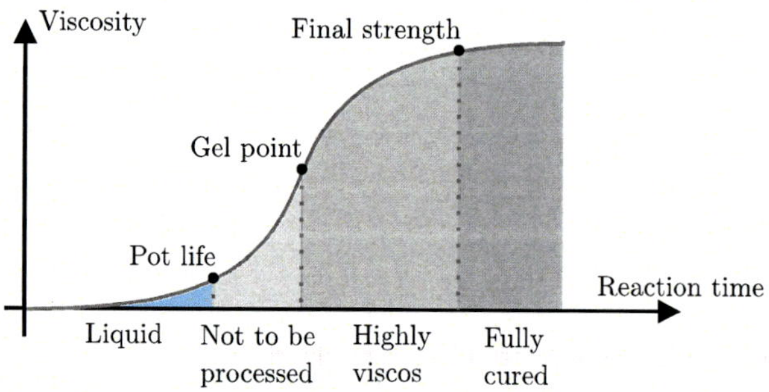

Fig. 9.3 Change of viscosity during chemical reaction

9.1.5 Viscosity

The effect of cross-linking is also reflected in a change of the viscosity. During the transition from liquid to solid, the viscosity increases more or less abruptly. The time when the drastic increase takes place is called the gel point (Fig. 9.3). The higher the process temperature, the earlier the gel point is reached. On the other hand, the viscosity decreases as the temperature increases.

9.2 Experimental Observations

Since 3D printing of biocompatible silicone was not possible with conventional methods, a new technology was developed in Stieghorst (2017). For the optimal design, some restraints have to be taken into account. To achieve a high degree of efficiency, the laser should allow high energy absorption. With silicone rubber, almost 90% of the energy is absorbed at a laser wavelength of 9.9 μm. The laser power must not be too high to prevent the thermal degradation of the material. The critical temperature for silicone rubber is 355 °C. If the temperature is close to this limit, the water in the silicone can evaporate.

The adjustable process parameters are print head velocity, extrusion velocity, laser power, nozzle diameter, initial viscosity, and substrate material. The following experimental observations are based on the investigations in Stieghorst (2017).

9.2.1 Drop Radius

The effects of substrate material, laser power, and initial viscosity on the spreading behavior of silicone can be determined by a simple experiment. A drop is deposited on a substrate, irradiated by a laser, and the maximum spread is measured. In addition to biocompatibility, surface tension is particularly important when selecting the substrate material. Depending on the material of the substrate, an increase of the initial radius between 25% and 175% can be observed.

The laser power also reduces the final radius. With Parylence C as substrate material, an increase of the power to 4.85 W/cm^2 leads to a spread reduction of 120%. When increased to 7.85 W/cm^2, no further significant reductions can be observed for all substrate materials.

A high initial viscosity leads to a faster cross-linking of the material. This also reduces the initial radius. However, an outflow of the liquid through the nozzle must be guaranteed. The maximum admissible initial viscosity depends on the dead volume, the pot life, and the extrusion velocity.

Furthermore, the short penetration depth of the laser into the material must be taken into account. At high temperatures, the outer layer is cross-linked first. This can cause thermal residual stresses in the material.

9.2.2 *Printing Shape*

During Additive Manufacturing, the print head velocity, the diameter of the nozzle, and the extrusion velocity also influence the geometry of the printed part. Increasing the print head velocity leads to a reduction in both line width and height. This relationship can be assumed to be linear. The maximum velocity depends on the design of the printer. Vibrations in the print head at too high velocities result in an undulating geometry of the printed layer.

With the same extrusion and print head velocity, the comparison of two different nozzle diameters (0.2 and 0.25 mm) shows no major differences in the minimum width and height of a layer. However, a large nozzle diameter allows a higher volume flow, and wider and higher layers can be achieved.

A further phenomenon can be observed during printing. If each layer is excited to an accelerated cross-linking, a stronger waviness occurs in the geometry than if only every second layer is irradiated.

9.3 Laser Modeling

In this 3D-printing process, a laser is selected whose beam has a doughnut shape in its cross-section (Fig. 9.4). The dispenser is in the dead center of the laser. The intensity distribution over the cross-section can be described by a normalized radial Gaussian function depending on the inner r_i and the outer radius r_a

$$I_{rad}\,(r) = I_0 \exp\left(-18\frac{(r-0.5r_i-0.5r_a)^2}{(0.5r_a-0.5r_i)^2}\right), \quad I_0 = \frac{3\sqrt{2}}{\pi^{1.5}(r_a-r_i)(r_a+r_i)\mathrm{erf}(3/\sqrt{2})}. \tag{9.2}$$

In this approach, the laser power is modeled as a heat flow at the surface

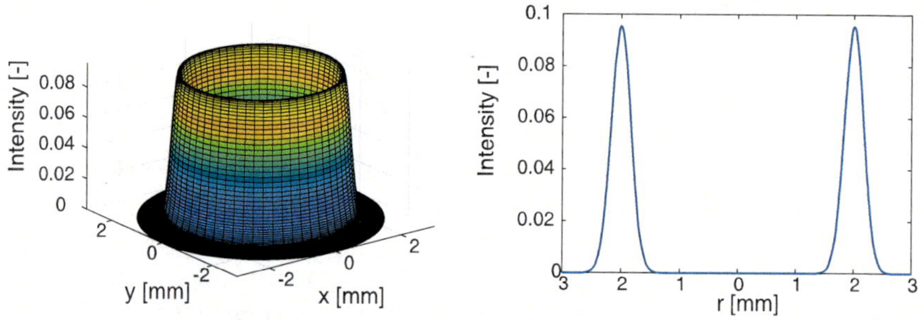

Fig. 9.4 (Left) Laser intensity distribution in the cross-section. (Right) Laser intensity distribution in radial direction (Hartmann et al. 2021)

$$\int_{\partial B_t^{N\Theta}} P_{laser} I_{rad} \, \mathrm{d}a = \int_{\partial B_t^{N\Theta}} \mathbf{q} \cdot \mathbf{n} \, \mathrm{d}a. \tag{9.3}$$

9.4 Modeling of Material Response

The modeling of Direct Poly Printing is based on the coupling of mechanical, thermal, and chemical differential equations. The balance of momentum forms the basis for the mechanical response behavior. The 1st law of thermodynamics provides the temperature distribution in the material. The modeling of cross-linking is based on the evolution equation for the degree of cure α. This quantity is coupled to the mechanical and thermodynamic differential equations via the specific strain energy function. The peridynamic concept based on the total Lagrangian description is used for the discretization and the balance of momentum formulated with respect to the initial configuration

$$\mathrm{Div}\,\mathbf{P}(\mathbf{F}, \vartheta, \alpha) + \rho_0 \bar{\mathbf{b}} = \rho_0 \ddot{\mathbf{u}}. \tag{9.4}$$

The divergence operator with respect to $\mathbf{X}$, the initial density ρ_0, and the 1st Piola Kirchhoff stress tensor $\mathbf{P}$ are given in (9.4). In the 1st law of thermodynamics, the influence of the degree of cure must be considered. Neglecting the Gough–Joule effect (3.50) formulated with respect to the initial configuration simplifies to

$$\rho_0 c_{\mathbf{F}} \dot{\vartheta} - \rho_0 \frac{\partial^2 \psi(\mathbf{F}, \vartheta, \alpha)}{\partial \vartheta \partial \alpha} \dot{\alpha} = D_{int} - \mathrm{Div}\,\mathbf{Q} + \rho_0 r. \tag{9.5}$$

To be consistent the Piola Kirchhoff heat flux vector $\mathbf{Q}$ must be used in (9.5). The temperature difference between the absolute Θ and initial temperature Θ_0 is chosen as the primary variable, i.e. $\vartheta = \Theta - \Theta_0$. In the unloaded initial state at ambient temperature, $\psi = 0$ must be valid. To ensure this condition, it is suitable to choose the temperature difference as an argument of the specific strain energy function.

9.4.1 Mechanical Constitutive Equations

In addition to mechanical loads, the effects of temperature and chemical reactions can lead to large deformations of the body. The deformation gradient is split into mechanical, thermal, and chemical parts. Since the temperature and the cross-linking result to a volumetric change of the body, the latter are described by the Jacobi determinant

$$\mathbf{F} = \mathbf{F}_M \mathbf{F}_\Theta \mathbf{F}_C, \quad \mathbf{F}_\Theta = J_\Theta^{\frac{1}{3}} \mathbf{1}, \quad \mathbf{F}_C = J_C^{\frac{1}{3}} \mathbf{1}. \tag{9.6}$$

In Lion and Johlitz (2012) and Landgraf et al. (2014) an approach for curing under large deformations based on a Maxwell model is presented. It can be classified as a finite linear viscoelastic model. To reproduce the flow behavior of the silicone more accurately, this formulation is extended by a friction element (Hartmann et al. 2021), see also Fig. 9.5. The necessity can be illustrated by a simple example (Hartmann 2019). For this purpose, a cylinder with a diameter of 0.25 mm and a depth of 0.5 mm is placed on a plate. The cylinder adheres to the substrate. The Young's modulus of the upper spring is $E_\infty = 15$ N/mm^2 and the Poisson ration $\nu = 0.499$. The shear modulus and viscosity in the lower branch is $\mu = 0.05$ N/m^2 and $\eta = 0.35$ Ns/m^2. If the purely viscoelastic approach is used, oscillations occur when the cylinder is deposited. But, the material does not spread which is unphysical. This only happens by adding a friction element ($Y = 0.25$ N/m^2) to the Maxwell curing model (Fig. 9.6).

Hence, the mechanical deformation gradient $\mathbf{F}_M$ is further split into an elastic, viscous and plastic part (Hartmann et al. 2021). To keep the number of indices as

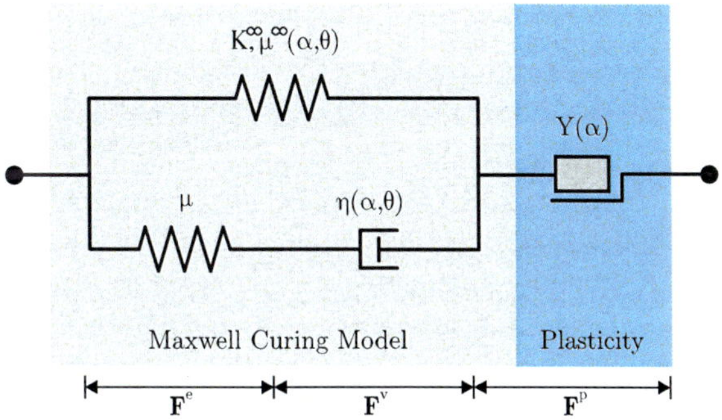

Fig. 9.5 Rheological model of silicone. Maxwell curing model with an additional friction element

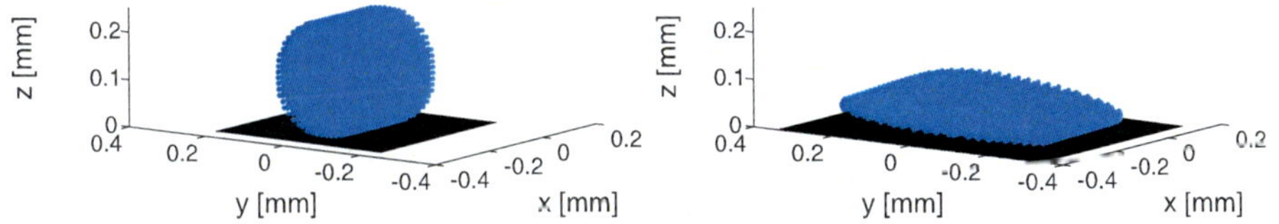

Fig. 9.6 Spreading behavior of a viscous cylinder. Situation at t = 0,02 s. (Left) Original Maxwell curing model. (Right) Curing model with additional friction element

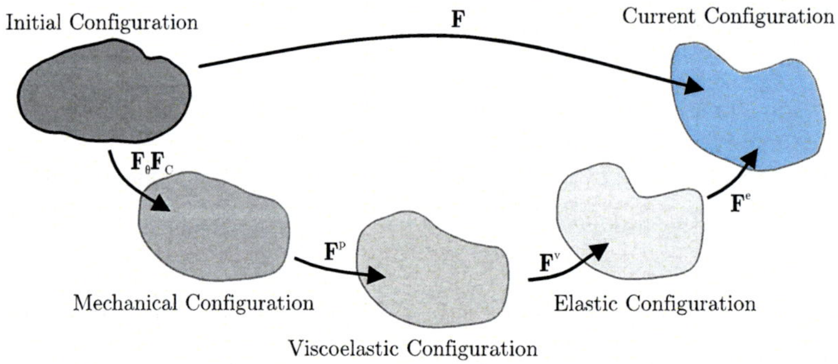

Fig. 9.7 Definition of three intermediate configurations in modeling curing of polymers

low as possible, the index M is omitted in the following, since thermal and chemical influences are only described via the Jacobi determinant

$$\mathbf{F}_M = \mathbf{F}^{ve}\mathbf{F}^p = \mathbf{F}^e\mathbf{F}^v\mathbf{F}^p. \tag{9.7}$$

The modeling of the stress is based on the right Cauchy Green tensor **C**. The viscoelastic component in the Maxwell model results from a push-forward of **C** to the corresponding intermediate configuration (Fig. 9.7)

$$\widetilde{\mathbf{C}}^{ve} = \mathbf{F}^{ve\,\mathrm{T}}\mathbf{F}^{ve} = J_\Theta^{-\frac{2}{3}} J_C^{-\frac{2}{3}} \mathbf{F}^{p\,-\mathrm{T}}\mathbf{C}\mathbf{F}^{p,-1}, \quad \mathbf{C} = \mathbf{F}^\mathrm{T}\mathbf{F}. \tag{9.8}$$

The tilde indicates quantities defined on this intermediate configuration. The Cauchy Green tensor in the lower spring results from the push-forward of **C** to the intermediate elastic configuration. Quantities with respect to this configuration are marked with a bar

$$\bar{\mathbf{C}}^e = \mathbf{F}^{e\,\mathrm{T}}\mathbf{F}^e = \mathbf{F}^{v\,-\mathrm{T}}\widetilde{\mathbf{C}}^{ve}\mathbf{F}^{v\,-1}. \tag{9.9}$$

In both springs a specific strain energy function can be defined in dependence on the corresponding kinematical measures. In the approach according to Lion and Höfer (2007) and Landgraf et al. (2014) an additional free thermochemical part $\psi_{\vartheta\alpha}(\vartheta, \alpha)$ is added

$$\psi := \psi_\infty(\widetilde{\mathbf{C}}^{ve}, \vartheta, \alpha) + \psi_\mu(\bar{\mathbf{C}}^e_{iso}) + \psi_{\Theta\alpha}(\vartheta, \alpha). \tag{9.10}$$

The stress in the lower spring is purely isochoric. The first term on the right-hand side in (9.10) is further subdivided into an isochoric and a volumetric part. The dependence on the temperature and degree of cure is only given in the isochoric part

$$\psi^\infty(\widetilde{\mathbf{C}}^e, \vartheta, \alpha) = \psi^{iso}_\infty(\widetilde{\mathbf{C}}^{ve}_{iso}, \vartheta, \alpha) + \psi^{vol}_\infty(J^{ve}). \tag{9.11}$$

To ensure that a change in crosslinking does not lead to a change in stress (see also Sect. 3.4), Lion and Johlitz (2012) introduces a pseudo-elastic approach for the isochoric part of the upper spring

$$\psi_{iso}^{\infty}(\widetilde{\mathbf{C}}_{iso}^{ve}, \vartheta) = \frac{1}{2} \int_{-\infty}^{t} \mu_{\infty}(s) \left(\frac{\mathrm{d}}{\mathrm{d}s} \widetilde{\mathbf{C}}_{iso}^{ve\,-1}(s) \right) \mathrm{d}s. \tag{9.12}$$

The 1st Piola Kirchhoff stress tensor results form the derivative of the specific strain energy function with respect to $\mathbf{C}$

$$\mathbf{P} = \mathbf{F}\mathbf{S} = 2\mathbf{F}\rho_0 \left[\frac{\partial \psi_{\infty}(\widetilde{\mathbf{C}}^{ve}, \vartheta, \alpha)}{\partial \mathbf{C}} + \frac{\partial \psi_{\mu}(\bar{\mathbf{C}}_{iso}^{e})}{\partial \mathbf{C}} \right]. \tag{9.13}$$

A detailed description of the derivative can be found in Hartmann (2019). The still unknown chemical and thermal parts of the Jacobi determinant are directly related to the solution functions ϑ and α by the material parameters β_C and β_Θ

$$J_\Theta = 1 + \beta_\Theta \vartheta, \quad J_C = 1 + \beta_C \alpha. \tag{9.14}$$

The material parameters in the Maxwell model depend on the temperature and the degree of cure. This applies on the one hand to the shear modulus in the upper spring and the dynamic viscosity

$$\begin{aligned} \mu(\Theta, \alpha) &:= f(\Theta) g(\theta) \mu_0, \quad f(\Theta) = \left(\frac{\Theta_0}{\Theta} \right)^{p_\Theta^\mu}, \; g(\alpha) = p_1^\mu (1 - \alpha) + p_2^\mu \alpha \\ \eta(\Theta, \alpha) &:= f(\Theta) g(\theta) \eta_0, \quad f(\Theta) = \left(\frac{\Theta_0}{\Theta} \right)^{p_\Theta^\eta}, \quad g(\alpha) = p_1^\eta (1 - \alpha) + p_2^\eta \alpha. \end{aligned} \tag{9.15}$$

The selected formulations are based on the approach described in Hossain and Steinmann (2014). The unknown quantities $\mathbf{F}^p$ and $\mathbf{F}^v$ are calculated from evolution equations. The standard Maxwell model is used for the viscous part

$$\frac{\mathrm{d}}{\mathrm{d}t} \left(\mathbf{F}^{v\,\mathrm{T}} \mathbf{F}^v \right) = \frac{1}{\tau} \left[\widetilde{\mathbf{C}}^{ve} - \mathbf{F}^{v\,\mathrm{T}} \mathbf{F}^v \right], \quad \tau = \frac{\eta}{\mu}. \tag{9.16}$$

The plastic deformation gradient is based on the von Mises model. To reproduce the transition from liquid to viscoelastic polymer, the limit stress Y now depends on the degree of cure (Hartmann et al. 2021). When the gel point α_{gel} is reached, the material behavior is assumed to be purely visoelastic

$$Y(\alpha) := \begin{cases} Y_0 + (Y_{gel} - Y_0) \frac{\alpha}{\alpha_{gel}} & \text{if } \alpha \leq \alpha_{gel} \\ \infty & \text{if } \alpha > \alpha_{gel} \end{cases}. \tag{9.17}$$

The evolution equation for plastic flow is formulated with respect to the current configuration

$$\mathbf{d}^p = \text{sym}\left[\mathbf{F}^{ve}\dot{\mathbf{F}}^p\mathbf{F}^{p\,-1}\mathbf{F}^{ve\,-1}\right] = \dot{\lambda}\frac{\partial f}{\partial \mathbf{s}}, \qquad f = \|\mathbf{s}\| - \sqrt{\frac{2}{3}}Y(\alpha). \tag{9.18}$$

An detailed derivation of the relationship between plastic symmetric velocity gradient $\mathbf{d}^p$ and plastic deformation gradient can be found e.g. in Wriggers (2008). The distinction between plastic and elastic state results from the Kuhn-Tucker condition $\lambda \geq 0$, $f \leq 0$ and $\lambda\, f = 0$. The yield function f depends on the deviatoric part of the Cauchy stress tensor $\boldsymbol{\sigma}$

$$\mathbf{s} = \boldsymbol{\sigma} - \frac{1}{3}\,(\text{tr}\,\boldsymbol{\sigma})\,\mathbf{1}, \qquad \boldsymbol{\sigma} = \frac{1}{J}\mathbf{P}\mathbf{F}^{\mathrm{T}}. \tag{9.19}$$

Details on plastic modeling and a numerical implementation can be found e.g. in Simo (1998).

9.4.2 Thermal Constitutive Equations

The specific entropy in the 1st law of thermodynamics computes from the derivative of the specific strain energy function (9.10) with respect to temperature

$$\eta = -\frac{\partial \psi}{\partial \vartheta} = -\frac{\partial \psi_\infty(\tilde{\mathbf{C}}^{ve}, \vartheta)}{\partial \Theta} - \frac{\partial \psi_{\vartheta\alpha}}{\partial \vartheta}. \tag{9.20}$$

The derivative of entropy with respect to temperature corresponds to the heat capacity. In Landgraf et al. (2014), the corresponding portion from the upper spring is neglected and the specific heat capacity depends solely on the specific thermochemical strain energy function

$$c_{\mathbf{F}}(\Theta, \alpha) := -\Theta\frac{\partial^2 \psi_{\Theta\alpha}}{\partial^2 \vartheta}. \tag{9.21}$$

The specific entropy is a function of the degree of cure that leads to an additional term in the 1st law of thermodynamics

$$\rho_0\, c_{\mathbf{F}}\dot{\vartheta} - \rho_0\Theta\frac{\partial^2 \psi_{\Theta\alpha}}{\partial \vartheta \partial \alpha}\dot{\alpha} = D_{int} - \text{Div}\,\mathbf{Q} + \rho_0\, r$$

The second term on the left-hand side describes the heat required to increase the degree of cure by one at constant temperature and deformation. The dissipative part contains the exothermic heat due to chemical reactions and the heat loss due to plasticity

$$D_{int} = -\rho_0 \frac{\partial \psi_{\Theta\alpha}}{\partial \alpha} \dot{\alpha} + \widetilde{\mathbf{C}}^{ve} \widetilde{\mathbf{S}} \cdot \widetilde{\mathbf{L}}^p. \tag{9.22}$$

The stress $\widetilde{\mathbf{S}}$ results from the push-forward of the 2nd Piola-Kirchhoff stress tensor to the viscoelastic configuration

$$\widetilde{\mathbf{S}} = J_{\Theta}^{\frac{2}{3}} J_C^{\frac{2}{3}} \mathbf{F}^p \mathbf{S} \mathbf{F}^{p,\mathrm{T}}, \quad \widetilde{\mathbf{L}}^p = \dot{\mathbf{F}}^p \mathbf{F}^{p\,-1} \tag{9.23}$$

In (9.5), the Gough–Joule effect is neglected since this contribution can be assumed to be very small (Boley and Weiner 1960).

Remark From the 2nd law of thermodynamics (3.24) additional parts result from the change of J_Θ and J_C in time. However, these are neglected, since their contributions can also be assumed to be small.

9.4.3 *Chemical Constitutive Equations*

The curing process can be modeled by a differential equation. Different approaches are available depending on the reaction behavior (Kiasat 2000). For the 3D-printing of silicone the model according to Sourour and Kamal (1976) is employed

$$\dot{\alpha} = \left[A_1 \exp\left(-\frac{B_1}{\Theta}\right) + A_2 \exp\left(-\frac{B_2}{\Theta}\right) \alpha^m \right] (1-\alpha)^n. \tag{9.24}$$

This description is based on six material parameters A_1, B_1, A_2, B_2, m, n, which have to be determined by suitable experiments.

9.5 Discretization Using Peridynamics

The discretization of the mechanical differential equation leads to an equilibrium of forces that must be fulfilled at each individual particle

$$\mathbf{f}_I^s + \mathbf{f}_I^g + \mathbf{f}_I^t = \begin{cases} \mathbf{f}_I^n & \text{boundary particle} \\ 0 & \text{inner particle} \end{cases}. \tag{9.25}$$

The contributions from the dead load $\mathbf{f}_I^g$ and the inertia term $\mathbf{f}_I^t$ are based only on quantities at the corresponding particle

$$\mathbf{f}_I^g = -m_I \bar{\mathbf{b}}, \quad \mathbf{f}_I^t = m_I \ddot{\mathbf{u}}_I. \tag{9.26}$$

To avoid negative effects due to under-integration the neighborhood is further subdivided based on the approach in Sect. 7.7. The internal force due to the state of stress in the material computes analogously to (7.39)

$$\mathbf{f}_I^s = -\sum_{\alpha=1}^{N_I^{sub}} \sum_{J=1}^{N_I^\alpha} \left[\mathbf{P}_I^\alpha \frac{\partial W\left(\mathbf{X}_J - \mathbf{X}_I\right)}{\partial \mathbf{X}_J} \left(\mathbf{K}_I^\alpha\right)^{-1} - \mathbf{P}_J^\alpha \frac{\partial W\left(\mathbf{X}_I - \mathbf{X}_J\right)}{\partial \mathbf{X}_I} \left(\mathbf{K}_J^\alpha\right)^{-1} \right] V_I V_J .$$

The 1st Piola Kirchhoff stress tensor is a function of the deformation gradient, the temperature, and the degree of cure at the corresponding particle $\mathbf{P}_I^a := \mathbf{f}(\mathbf{F}_I^a, \Theta_I, \alpha_I)$. The deformation gradient is determined analogously to (7.26), whereby the evaluation only takes place over the subhorizon

$$\mathbf{F}_I^\alpha = \sum_{J=1}^{N_I^a} \left(\mathbf{x}_J - \mathbf{x}_I\right) \otimes \frac{\partial W\left(\mathbf{X}_J - \mathbf{X}_I\right)}{\partial \mathbf{X}_J} \left(\mathbf{K}_I^\alpha\right)^{-1} V_J . \tag{9.27}$$

The stress is calculated from (9.13). To ease the notation, the upper index a, indicating the evaluation at the subhorizon, is neglected. The kinematic quantities $\widetilde{\mathbf{C}}^{ve}$ and $\bar{\mathbf{C}}^e$ result from the deformation gradients $\mathbf{F}_I$, $\mathbf{F}_I^p$, and $\mathbf{F}_I^v$ as well as the thermal and chemical Jacobi determinantss. The plastic and viscous components are calculated by integrating the evolution Eqs. (9.16) and (9.18) employing the implicit Euler method. The viscous deformation gradient is determined from

$$\mathbf{F}^{v\,\mathrm{T}}_{I\,n+1}\mathbf{F}^{v}_{I\,n+1} = \left(1 + \frac{\Delta t}{\tau_{I\,n+1}}\right)^{-1} \left[\mathbf{F}^{v\,\mathrm{T}}_{I\,n}\mathbf{F}^{v}_{I\,n} + \frac{\Delta t}{\tau_{I\,n+1}} \widetilde{\mathbf{C}}^{ve}_{I\,n+1}\right] . \tag{9.28}$$

The exponential map integrator is applied to integrate the plastic evolution equation. Exploiting the symmetry of the direction vector $\mathbf{n}$, the current plastic deformation gradient results from

$$\mathbf{F}^{p}_{I\,n+1} = \exp\!\left(\Delta\lambda_I \mathbf{F}^{ve\,-1}_{I\,n+1} \mathbf{n}_{I\,n+1} \mathbf{F}^{ve}_{I\,n+1}\right) \mathbf{F}^{p}_{I\,n}, \quad \mathbf{n}_{I\,n+1} = \frac{\partial f_{I\,n+1}}{\partial \mathbf{s}} . \tag{9.29}$$

An exact derivation of the above equation in the context of FEM can be found in Korelc and Stupkiewicz (2014). The Newton method is employed to solve the unknown variables $\mathbf{F}_I^p$, $\mathbf{F}_I^v$ and $\Delta\lambda_I$. The derivations contained therein are generated with the software package AceGen (Korelc and Wriggers 2016). The influence of surface tension is not included in the simulation but can be integrated into the algorithm analogously to Sect. 8.5.1.

The temperature at a particle result from the solution of the thermal differential equation. In the discrete form, the powers at each particle must be in equilibrium

$$P_I^\eta + P_I^d + P_I^q = \begin{cases} P_I^n & \text{boundary particle} \\ 0 & \text{inner particle} \end{cases} . \tag{9.30}$$

The term P_I^d due to thermal conduction is computed using a bond based approach introduced in Oterkus et al. (2014). The corresponding power is determined solely by the temperature difference between particles

$$P_I^q = V_I \sum_{J=1}^{N_I} g_{IJ} \kappa_I \frac{\Theta_J - \Theta_I}{\|\mathbf{X}_J - \mathbf{X}_I\|} V_J. \tag{9.31}$$

The heat conduction κ_I depends on the size of the neighborhood and is calculated from the heat conduction coefficient k and the horizon δ_I. In bond-based formulations, the reproducing conditions are not automatically fulfilled, especially at the boundary. Therefore, the correction term g_{IJ} is introduced. This term describes the ratio of the maximum volume of all neighborhoods to the current volume of the particle (Hartmann et al. 2021)

$$\kappa_I = \frac{6k}{\delta_I^4 \pi}, \quad g_{IJ} = \frac{1}{2}\left(g_I + g_J\right), \quad g_I = \frac{\bar{V}_{max}}{V_I}, \quad \bar{V}_I = \sum_{J=1}^{N_I} V_J, \quad \bar{V}_{max} = \max_I \bar{V}_I. \tag{9.32}$$

The entropy and dissipation components are determined directly from the corresponding values at the particle

$$P_I^\eta = -m_I\, c_I \dot{\vartheta}_I + m_I \Theta_I \frac{\partial^2 \psi_{\Theta\alpha}}{\partial\Theta\partial\alpha} \dot{\alpha}_I, \quad P_I^d = -m_I \frac{\partial \psi_{\Theta\alpha}}{\partial\alpha} \dot{\alpha}_I + V_I \widetilde{\mathbf{C}}_I^{ve} \widetilde{\mathbf{S}}_I \cdot \widetilde{\mathbf{L}}_I^p. \tag{9.33}$$

The power from the laser is absorbed by the boundary particles, which are irradiated by the laser. At the remaining particles at the surface, heat exchange with the environment takes place which computes from the ambient temperature which is equal to the ambient temperature Θ_0 and the transition coefficient h

$$P_I^n := \begin{cases} P_{laser} I\left(\mathbf{x}_I^s\right) A_I & \text{beneath the laser} \\ h\left(\Theta_0 - \Theta_I\right) A_I & \text{contact with environment} \end{cases}. \tag{9.34}$$

This contribution requires the surface area A_I. The first step is the identification of the boundary particles. The integration constraint (7.35) can be used for this purpose. Each particle which possesses a non-zero resulting normal vector $\bar{\mathbf{N}}_I$ is treated as a boundary particle. The norm provides the corresponding area

$$\begin{aligned} A_I = \|\bar{\mathbf{N}}_I\| = \Bigg\| \sum_{J=1}^{N_I} \Big[& \omega\left(\mathbf{X}_J - \mathbf{X}_I\right) \mathbf{K}_I^{-1} \left(\mathbf{X}_J - \mathbf{X}_I\right) \\ & - \omega\left(\mathbf{X}_I - \mathbf{X}_J\right) \mathbf{K}_J^{-1} \left(\mathbf{X}_I - \mathbf{X}_J\right) \Big] V_I V_J \Bigg\|. \end{aligned} \tag{9.35}$$

The degree of cure of a particle results directly from the discretization of the evolution Eq. (9.24)

$$\dot{\alpha}_I = \left[A_1 \exp\left(-\frac{B_1}{\Theta_I}\right) + A_2 \exp\left(-\frac{B_2}{\Theta_I}\right)\alpha_I^m \right] (1 - \alpha_I)^n. \tag{9.36}$$

9.5.1 *Time Integration*

To determine the solution function, integration over time is required. Usually, explicit methods are applied in Peridynamics. The acceleration is integrated using the Verlet scheme. This method ensures an accurarcy of 2nd order

$$\mathbf{u}_{I\,n+1} = \mathbf{u}_{I\,n} + \mathbf{v}_{I\,n}\Delta t + \frac{1}{2}\ddot{\mathbf{u}}_{I\,n+1}(\Delta t)^2, \quad \mathbf{v}_{I\,n+1} = \mathbf{v}_{I\,n} + \frac{1}{2}\left(\ddot{\mathbf{u}}_{I\,n+1} + \ddot{\mathbf{u}}_{I\,n}\right)\Delta t. \tag{9.37}$$

The temperature and the degree of cure at the next time step are determined by the explicit Euler scheme

$$\Theta_{I\,n+1} = \Theta_{I\,n} + \Delta t \dot{\Theta}_{I\,n}, \quad \alpha_{I\,n+1} = \alpha_{I\,n} + \Delta t \dot{\alpha}_{I\,n}. \tag{9.38}$$

The critical time step results from the fulfillment of the criteria for the mechanical and thermal differential equation

$$\Delta t = \min\left(\frac{h}{c}, \frac{\rho_0\, c_0}{\bar{k}}\right), \qquad c = \sqrt{\frac{K}{\rho}}, \quad \bar{k} = \max_I \sum_{J=1}^{N_I} \frac{\kappa_J}{\|\mathbf{X}_J - \mathbf{X}_I\|}. \tag{9.39}$$

The second condition is taken from Madenci and Oterkus (2014).

9.6 Computational Studies

The presented examples investigate the influence of extrusion velocity and laser power on the deformation behavior of the deposited silicone. For this purpose, a representative 3D-printing step is simulated which reflects the characteristic features of the whole process (Hartmann et al. 2021). A part of the silicone is already deposited (Fig. 9.8). In all simulations, the diameter of the extruder nozzle is 0.25 mm and the print head moves with a velocity of 0.01 m/s. After approximately 0.05 s, the behavior of the printing process can be assumed to be stationary. This corresponds to a deposited layer of 0.75 mm. To fulfill the requirement (9.39) of explicit time integration schemes, a time step size of $0.5 \cdot 10^{-7}$ s is selected. The search radius of the neighborhood is $\delta = 3\,\|\Delta\mathbf{X}\|$. The distance between the particles in the initial configuration $\|\Delta\mathbf{X}\|$ is constant in the entire domain. The silicone sticks to the substrate. Thus, a particle adheres when it comes into contact with the substrate. The displacements are then fixed in all directions and a detachment from the plate is excluded. The characterization of the material behavior is determined by the specific

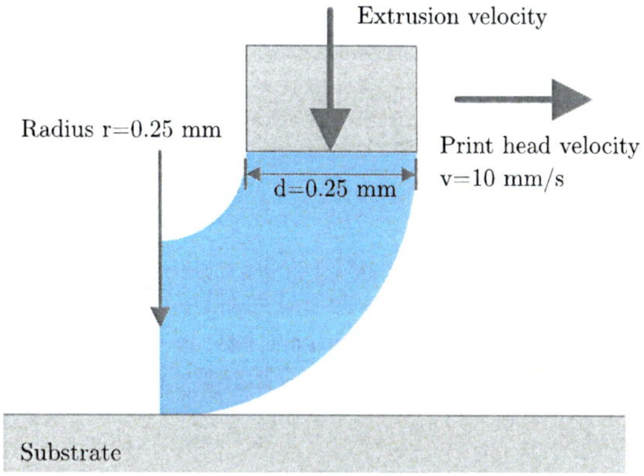

Fig. 9.8 Geometry and dimensions of a representative 3D Direct Poly Printing process

strain energy functions. A Neo-Hookian model is used leading to the specific strain energy functions in (9.10)

$$\psi_\infty^{vol} = \frac{K}{4}\left[\left(J^{ve}\right)^2 - 1\right] - \frac{K}{2}\ln(J^{ve}), \quad \psi_\mu = \frac{1}{2}\mu\left(\operatorname{tr}\bar{\mathbf{C}}^e_{iso} - 3\right). \tag{9.40}$$

The material parameters for the Maxwell model and the friction element is given in Table 9.1. Since the exact values for medical silicone are not known, the values are estimated using comparable materials. For their dependence on temperature and degree of cure the parameters in (9.14), (9.15), and (9.17) need to be specified, see Table 9.2. The viscosity is assumed to be constant. The thermal conduction coefficient κ is 0.27 W/(m^{2}K). The heat capacity is calculated by deriving the specific thermochemical strain energy function (9.21)

$$\psi_{\Theta\alpha} = h_{F0} + \Delta h_{FS}\alpha - \left(\frac{1}{2}a_F\vartheta^2 + \frac{1}{6}b_F\vartheta^3\right)(1-\alpha) - \left(\frac{1}{2}a_S\vartheta^2 + \frac{1}{6}b_S\vartheta^3\right)\alpha. \tag{9.41}$$

The corresponding material parameters can be found in Table 9.3. The evolution equation for the degree of cure (9.24) requires additional parameters, which are summarized in Table 9.4. The influence of wetting is not included in the modeling.

Table 9.1 Material parameters for the Maxwell model and the friction element

Parameter:	K^∞	μ_0^∞	μ	η	Y_0
Value:	20000	40	5.0	3.5	0.4
Unit:	N/m^2	N/m^2	N/m^2	Ns/m^2	N/m^2

Table 9.2 Additional parameters for temperature and degree of cure dependent variables

Parameter:	β_Θ	β_α	p_Θ	$p^\mu_{\alpha_1}$	$p^\mu_{\alpha_2}$	Y_{gel}	α_{gel}
Value:	10^{-5}	−0.03	5.0	5.0	613.75	1.2	0.1
Unit:	1/K	–	Pa	N/m^2	N/m^2	N/m^2	–

Table 9.3 Fitted material parameters for the definition of quantities based on a DSC experiment for acrylic bone cement (Landgraf 2015)

Parameter:	Δh_{FS}	a_F	b_F	a_S	b_S
Value:	$-80, 95$	$4, 5729\ 10^{-3}$	$-1, 9600\ 10^{-6}$	$4, 1210\ 10^{-3}$	$-1, 6785\ 10^{-6}$
Unit:	J/g	J/(gK^2)	J/(gK^3)	J/(gK^2)	J/(gK^3)

Table 9.4 Material parameters for the evolution equation of the degree of cure

Parameter:	A_1	A_2	B_1	B_2	m	n
Value:	60000.0	18000.0	25.0	30.5	2.7	1.6
Unit:	–	–	K	K	–	–

9.6.1 Influence of Extrusion Velocity

The first application investigates the effect of the extrusion velocity on the width of the printed layer. The test is performed at room temperature, i.e. 20 °C. The laser is switched off and the temperature in the body is assumed to be constant. Three different extrusion velocities, $v_e = 0.0025$ m/s, $v_e = 0.05$ m/s and $v_e = 0.01$ m/s are compared. As expected, an increase in velocity leads to more material being deposited per time unit, and hence, to an increase in width (Fig. 9.9). These qualitative results agree with the experimental observations in Stieghorst (2017). Figure 9.10 depicts

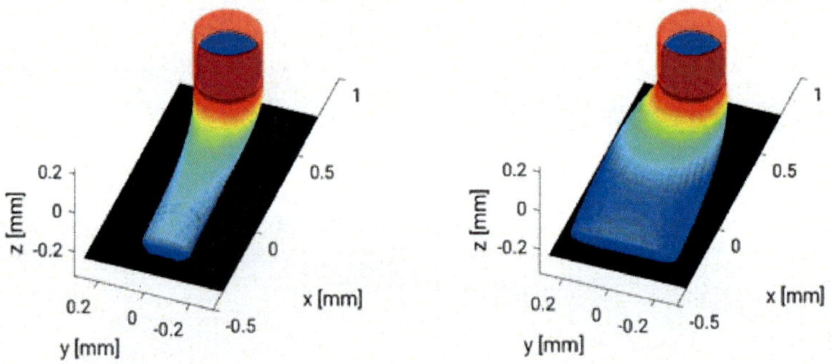

Fig. 9.9 Final shape of the printed layer. (Left) Extrusion velocity of 0.0025 m/s. (Right) Extrusion velocity of 0.01 m/s

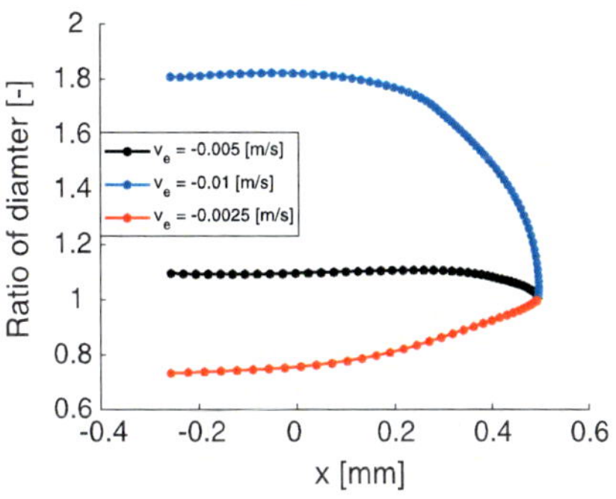

Fig. 9.10 Width of the layer to diameter of the nozzle of the final printed part along the printing direction

the change of width of the deposited layer along the printing direction. The maximum width changes from 0.18 to 0.28 mm and 0.45 mm if the velocity increases.

9.6.2 *Influence of Laser Power*

The second test examines the effect of heat input by the laser on the width of the printed layer. The laser power is 1.5 W. The wavelength allows 90% of the power to be absorbed by the material. The radii of the laser irradiation (9.2) are $r_i = 0.25$ mm and $r_a = 0.35$ mm. The extrusion velocity is set to $v_e = 0.01$ m/s. As can be seen in Fig. 9.11, the irradiation leads to a reduction of the width from 0.45 to 0.31 mm. The maximum temperature is 208.32 °C and below the critical value of 355 °C.

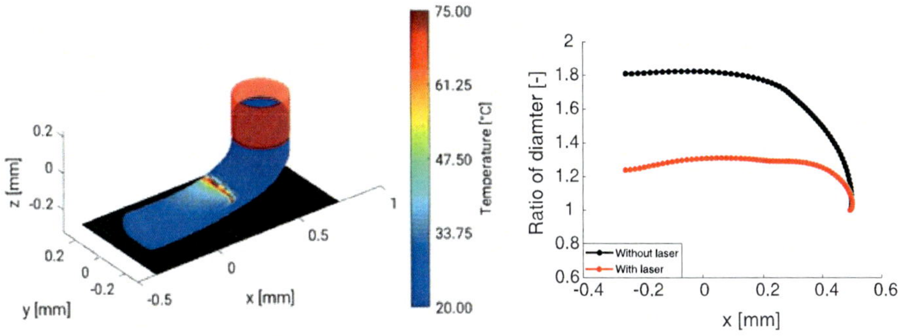

Fig. 9.11 (Left) Temperature distribution and shape of the final printed layer. (Right) Width of the layer to diameter of the nozzle of the final printed part along the printing direction

9.6.3 Conclusion

The examples only show the possibilities of a numerical investigation. A realistic simulation requires the exact specification of the material parameters for medical silicone. In addition, the influence of wetting must be taken into account.

References

B. Boley, J. Weiner, *Theory of Thermal Stresses* (Wiley, London, 1960)

H.G. Elias, *Polymere: Von Monomeren Und Makromolekülen Zu Werkstoffen; Eine Einführung* (UTB, Uni-Taschenbücher-GmbH, 1996)

P. Hartmann, Simulation of Thermo-Chemo-Mechanical Coupled Additive Manufacturing Processes using Peridynamics. Ph.D. thesis, Leibniz Universität Hannover, Germany (2019)

P. Hartmann, C. Weißenfels, P. Wriggers, A curing model for the numerical simulation within Additive Manufacturing of soft polymers using Peridynamics. Comput. Part. Mech. **8**(2), 369–388 (2021)

M. Hossain, P. Steinmann, Degree of cure-dependent modelling for polymer curing processes at small-strain. Part I: Consistent reformulation. Comput. Mech. **53**(4), 777–787 (2014)

M.S. Kiasat, Curing Shrinkage and Residual Stresses in Viscoelastic Thermosetting Resins and Composites. Ph.D. thesis, TU Delft, Netherlands (2000)

J. Korelc, P. Wriggers, *Automation of Finite Element Methods* (Springer, Berlin, 2016)

J. Korelc, S. Stupkiewicz, Closed-form matrix exponential and its application in finite-strain plasticity. In. J. Numer. Methods Eng. **98**(13), 960–987 (2014)

R. Landgraf, Modellierung und Simulation der Aushärtung polymerer Werkstoffe. Ph.D. thesis, Technische Universität Chemnitz, Germany (2015)

R. Landgraf, M. Rudolph, R. Scherzer, J. Ihlemann, Modelling and simulation of adhesive curing processes in bonded piezo metal composites. Comput. Mech. **54**(2), 547–565 (2014)

A. Lion, P. Höfer, On the phenomenological representation of curing phenomena in continuum mechanics. Arch. Mech. **59**(1), 59–89 (2007)

A. Lion, M. Johlitz, On the representation of chemical ageing of rubber in continuum mechanics. Int. J. Solids Struct. **49**(10), 1227–1240 (2012)

E. Madenci, E. Oterkus, *Peridynamic Theory and Its Applications* (Springer, Berlin, 2014)

S. Oterkus, E. Madenci, A. Agwai, Peridynamic thermal diffusion. J. Comput. Phys. **265**, 71–96 (2014)

J.C. Simo, *Numerical Analysis and Simulation of Plasticity*. Handbook of Numerical Analysis VI (Elsevier Science B.V., 1998)

S. Sourour, M.R. Kamal, Differential scanning calorimetry of epoxy cure: isothermal cure kinetics. Thermochimica Acta **14**(1–2), 41–59 (1976)

J. Stieghorst, Entwicklung eines additiven Fertigungsverfahrens für Silikonkautschuke zur Herstellung von flexiblen Neuroimplantaten in der Medizintechnik. Ph.D. thesis, Leibniz Universität Hannover, Germany (2017)

P. Wriggers, *Nonlinear Finite Element Methods* (Springer Science & Business Media, Berlin, 2008)